P9-CST-194

STAR MAPS

# STAR MAPS

**Astonishing new evidence from Ancient Civilisations and modern scientific research of man's origins and return to the stars**

**WM. R. FIX**

*A Jonathan-James Book*

OCTOPUS
MAYFLOWER

This edition was first published in 1979 by
Octopus Books Limited
59 Grosvenor Street
London W1



ISBN 0 7064 1085 8 (hardback)
ISBN 0 7064 1066 1 (paperback)

JONATHAN-JAMES BOOKS
5 Sultan Street
Toronto Ontario
Canada M5S 1L6

Editor: R. Carolyn King
Designer: Brant Cowie
Printed in the United States of America

*To Diannellen Starchild*

*'I am a child of earth and starry heaven, but my race is of heaven alone.'*

# CONTENTS


1 THE STRATAGEMS OF TIME 9

2 A 4000 YEAR OLD PICTURE 19

3 STAR TEMPLES 41

4 STAR MAPS 65

5 THE STAR WALKER 89

6 ON THE TRAIL OF UNAS 107

7 INITIATION 123

8 THE STAR SHIPS 159

9 THE RIVERS OF HEAVEN 169

10 THE GOLDEN PLATE OF PETELIA 185

11 THE LAKE OF MEMORY 203

12 THE GODS OF LIGHT 215

NOTES 233

WORKS CONSULTED 243

CHAPTER ONE

# THE STRATAGEMS OF TIME

## Tracing Man's Origins from Fragments of Ancient Civilisations

Nothing is more mysterious to man than his own origin; nothing stranger than the works and beliefs of his beginnings.

It is only within the last few decades that man has glimpsed the incredible antiquity of his tenure on earth. According to anthropologists who have traced man's beginnings by examining skeletal remains, the species of man to which we belong—*homo sapiens*—has been on this planet for over 100,000 years. Historians, on the other hand, are more concerned with man's works, such as documents, artifacts and architecture. Unfortunately for most areas of the earth, documented history reaches back only a few centuries; beyond that we have folklore, legend and myth. Even in Egypt, where the documents go back the farthest, we have only the vaguest conjectural ideas of history before 1600 BC. On the scale of man's existence as a species, it is as if our historical horizon extended back but three years out of a thousand.

The more that the age of *homo sapiens* has been pushed back, the more also have we found that beyond the 3500 year range of our historical vision there were earlier high civilizations that vanished in the thousand unknown centuries which preceded. Until recently contemporary scholars and educators have been virtually unanimous in defining history as an ascending line of progress from stone age barbarism to modern man, with his sciences and institutions as the most advanced attainments of the race —an idea easily accepted since it values the present above the past. We are accustomed to thinking that time has somehow played us the favourite. But the workings of time are not so simple.

One of the stratagems of time is the extraordinary juxtaposition of the oldest things in the world with the most recent. Researchers have uncovered increasing indications that we are not the first high scientific civilization on this planet. The more sophisticated we have become, the more we have found that many ancient remains, long considered the works and words of primitive men, reflect a similar but unexpected sophistication. As our own awareness has expanded, we find that many of the oldest utterances and remains of man serve to complement contemporary knowledge in a manner that is both intriguing and alarming.

Homer's Troy was nothing more than a mythical place in a poem until a century ago when Heinrich Schliemann followed geographic clues in the *Iliad* to locate the famous city in northwest Turkey. In *De Bello Gallico* Caesar wrote that the Druids discussed and imparted to the young 'many things concerning the heavenly bodies and their movements, the size of the world and of our earth, natural science, and the influence and power of the immortal gods'.[1]

Despite Caesar's words, until recently little credit was given to the Celts or any other ancient peoples as far as their knowledge of 'the heavenly bodies and their movements' was concerned. In the last few years Gerald Hawkins and other astronomers have shown that 4000 years ago Stonehenge functioned as a giant astronomic device allowing those who built it to predict eclipses and other celestial events about as well as we can today.[2] This seems to be the very knowledge alluded to by Caesar, which must have been preserved at least until the first century BC. Within the context of British prehistory, Alexander Thom has documented that many lesser stone circles in the British Isles were also used to observe significant risings and settings of celestial bodies.[3] Even more recently, remains of a similar ancient astronomic science have been detected in North America.[4] Somehow on both sides of the Atlantic, this knowledge was lost since Caesar's time.

Whatever ancient astronomic abilities may have been, historians have tended to be even more sceptical that anyone earlier than the Greeks could have been seriously concerned about 'the size of the world and of our

*Winter sunset at Stonehenge. By marking the extreme rising and setting positions of the sun and moon on the horizon with alignments of stones, the builders of Stonehenge obtained information for regulating the calendar and predicting the occurrence of eclipses.*

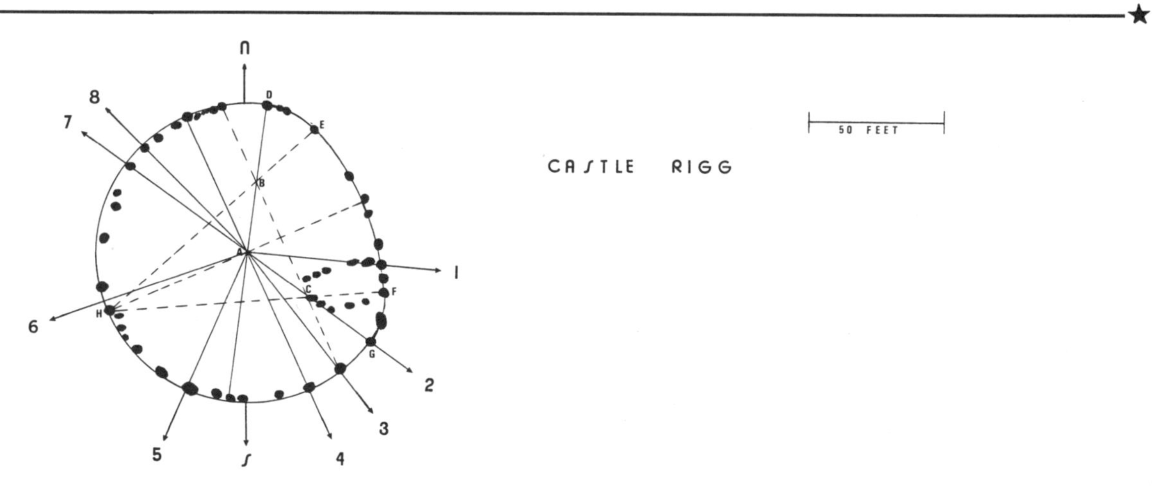


*The ground plan of a stone circle called Castle Rigg in Cumberland.*

*Sketch of Alexander Thom's survey with geometric construction superimposed. Arc DE is struck from B; Arc EF from H; Arc FG from C; Arc GHD from A. Note outline of right triangles. Thom refers to this construction as a 'flattened circle, Type A'.*

*The major declinations indicated are: 1. Equinox sunrise. 2. Candlemas rising sun. 3. Midwinter rising sun. 4. Most southerly rising moon. 5. To notch on the horizon and most southerly setting moon. 6. To outlying stone (90 metres, 296 feet). 7. Midsummer setting sun. 8. Most northerly setting moon.*

earth'. However, in *Maps of the Ancient Sea Kings* Charles Hapgood demonstrates that in deep antiquity—6000 to 8000 years ago and possibly earlier—someone recorded the subglacial topography of Antarctica.

As late as the early 16th century, some map makers were still using information recorded thousands of years earlier by a vanished civilization that had travelled all the earth. Some of these 16th century maps have survived until our day and have been intensively scrutinized by men like Hapgood and others. In 1531 a geographer named Oronteus Finaeus published a world map that showed the entire continent of Antarctica, which was not 'discovered' until 1818. Whoever drew the original source map from which Finaeus derived his information had detailed knowledge of the southern continent. Some of these details were in advance of 20th century cartography, until measurements taken during the International Geophysical Year, 1957-8, verified features shown on the Finaeus map which are now under mile-thick ice. Estimates of the age of this ice-sheet vary from 6000 to 100,000 years.

Hapgood's remarkable findings do not exist in isolation. In the last few years additional evidence has emerged which powerfully corroborates the existence of a highly developed science of geography in the prehistoric era. Through the study of ancient units of measure, historian L.C. Stecchini has detailed the nature of such a science of geography that measured distances in degrees, minutes and seconds of latitude and longitude, just as we do today.[5] In my previous book, *Pyramid Odyssey*, I have demonstrated how the builders of the Great Pyramid successfully measured the entire earth and recorded these measurements in the dimensions of the Great Pyramid on the scale of 1:43,200. It would seem that whoever built the Great Pyramid knew the earth's circumference and polar radius with astonishing precision, recording both the flattening at the poles and the equatorial bulge and they knew them with an accuracy comparable to that recorded by satellite surveys from space. Thus we find knowledge from the remote past that is disquieting. We are accustomed to thinking that we are ever progressing.

When Marco Polo returned to Venice around 1295 after some 25 years in Asia, he wrote a book in which he described what he had seen. The book was an instant success, spreading in a few months throughout Italy. However, people did not see his work as information on history, geography or travel but as a fantastic romance.

It is easy for us to look back at Polo's contemporaries and say that they were living in the Dark Ages. Seven hundred years from now it will probably be as easy for someone else to point at us in the same way. A more

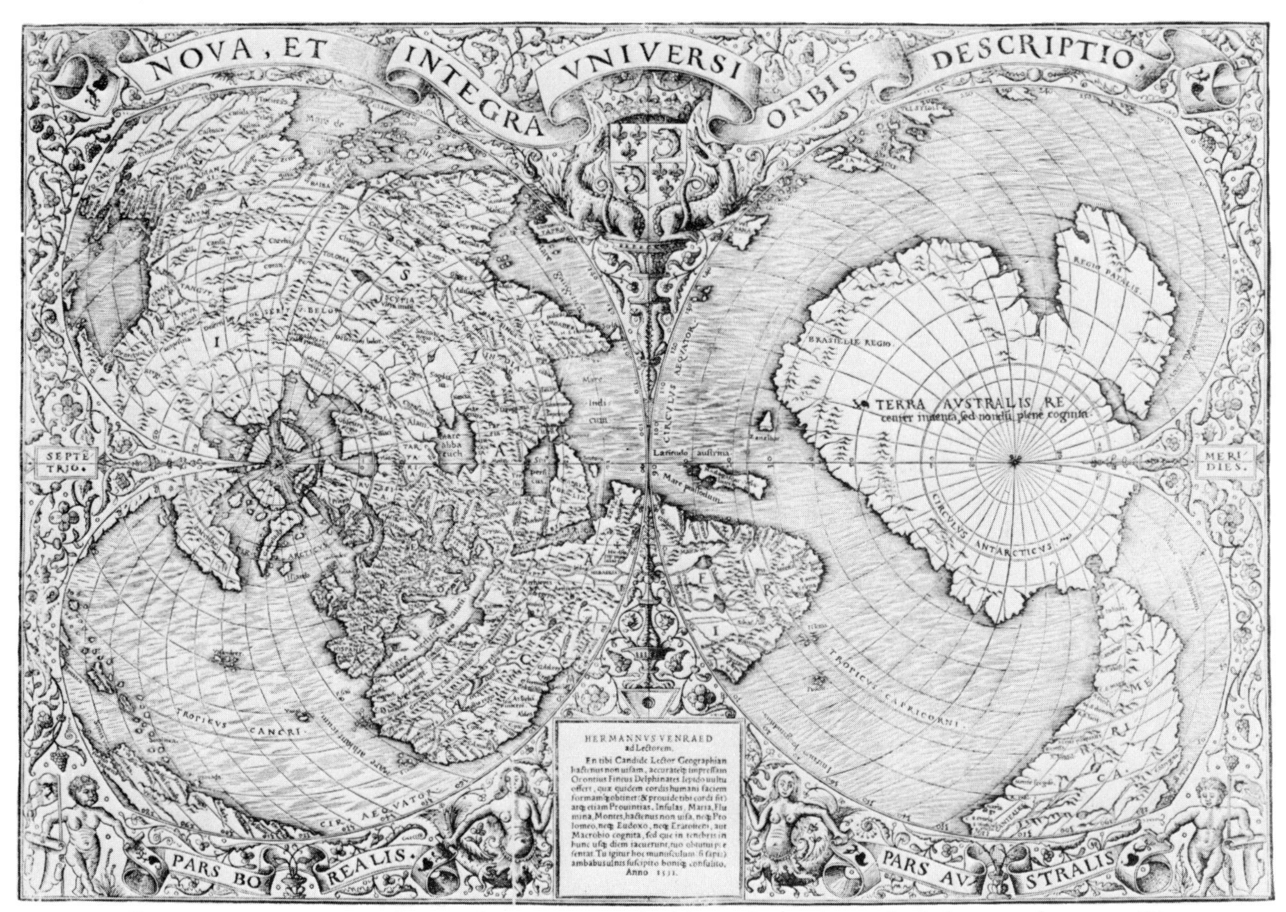


*The Oronteus Finaeus map of 1531. The greatest error in this map is that Antarctica is drawn too large, possibly a copyist's mistake, although mountains and other details, not rediscovered until 1958, are accurately represented.*

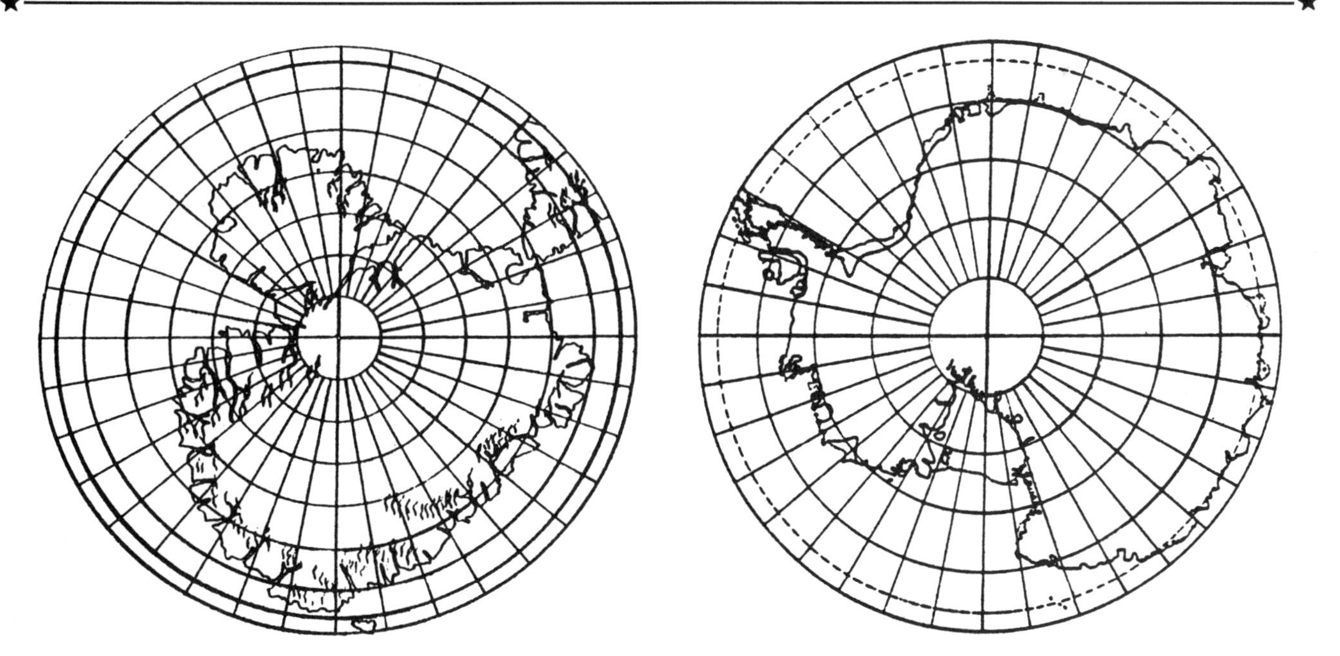

*Antarctica on the Oronteus Finaeus map of 1531 (left) reduced to the same scale and grid as a modern map of Antarctica.*

*The Great Pyramid at Gizeh. Note the figures at the base for scale.*

*Casing blocks on the north face of the Great Pyramid and the platform on which the Pyramid rests.*

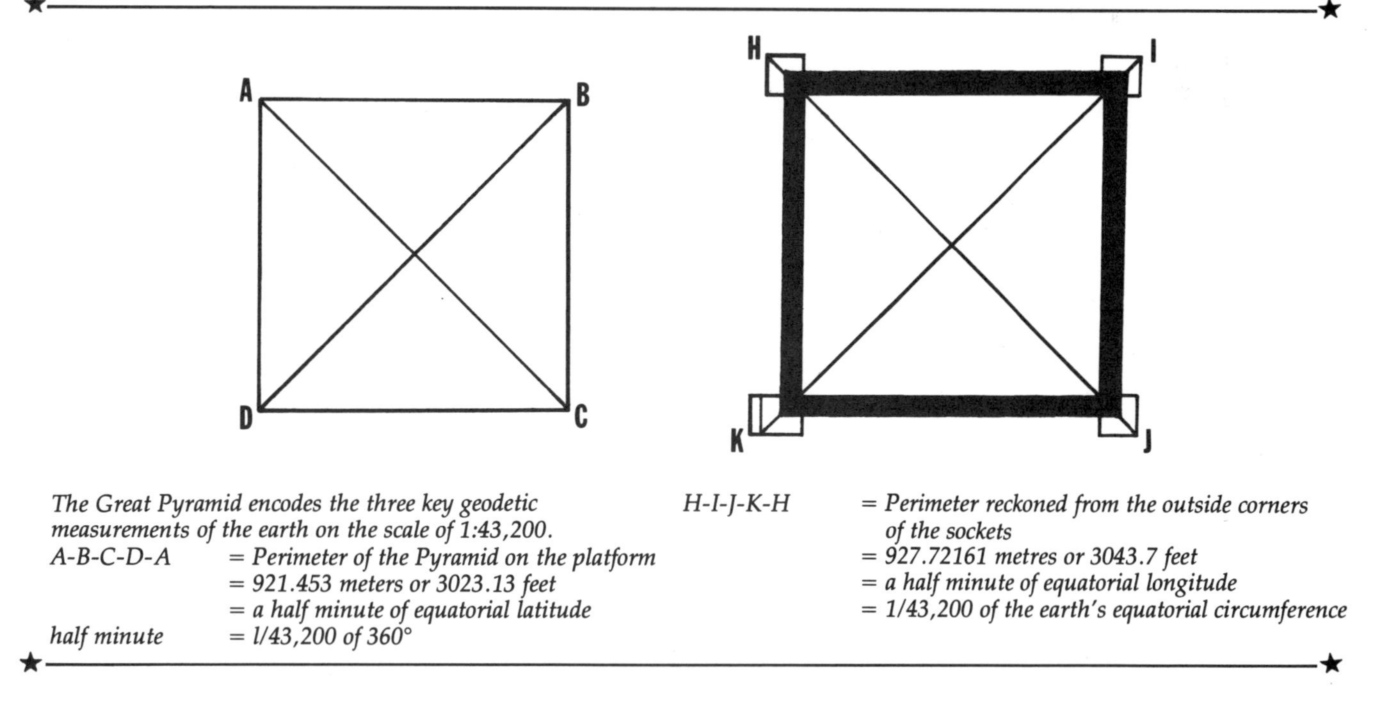


*The Great Pyramid encodes the three key geodetic measurements of the earth on the scale of 1:43,200.*

*A-B-C-D-A = Perimeter of the Pyramid on the platform*
*= 921.453 meters or 3023.13 feet*
*= a half minute of equatorial latitude*

*half minute = l/43,200 of 360°*

*H-I-J-K-H = Perimeter reckoned from the outside corners of the sockets*
*= 927.72161 metres or 3043.7 feet*
*= a half minute of equatorial longitude*
*= 1/43,200 of the earth's equatorial circumference*

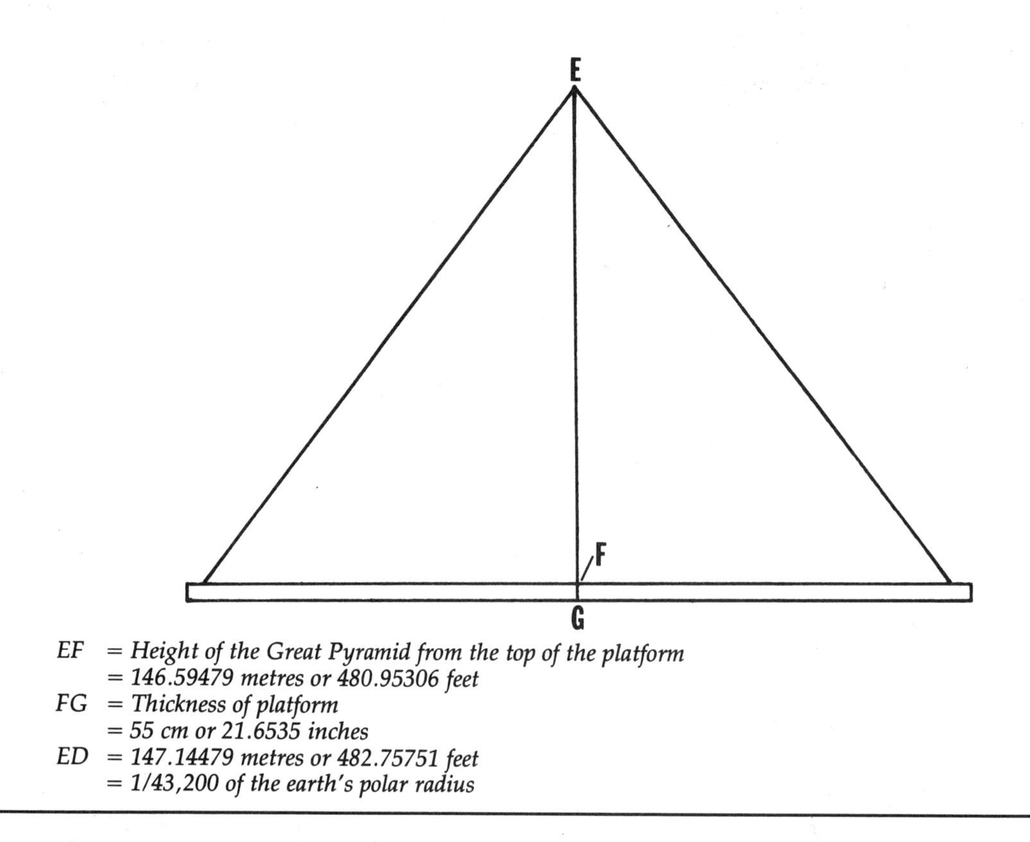


*EF = Height of the Great Pyramid from the top of the platform*
*= 146.59479 metres or 480.95306 feet*
*FG = Thickness of platform*
*= 55 cm or 21.6535 inches*
*ED = 147.14479 metres or 482.75751 feet*
*= 1/43,200 of the earth's polar radius*

alarming possibility is that if someone who lived 5000 years ago were brought into the 20th century, he might point at us as well as at Polo's contemporaries and declare that we were all living in the Dark Ages. An educated ancient Egyptian, for example, might have had answers to questions—very compelling answers—for which we are still searching today. Between his time and ours the great library of Alexandria, reputed to have once held 700,000 volumes and to have been the world's foremost centre of knowledge, was burned three times and has now completely disappeared. The loss of that and additional knowledge through countless other acts of destruction has diminished human awareness to the point that—by the Egyptian's standards—we hardly know who we are.

Yet as we have seen, the world of 5000 years ago has not completely vanished. Despite the incredible antiquity and intervening disasters wrought by man and nature, it is occasionally possible to glimpse the concerns of that ancient world. In the case of astronomy and geography these concerns now appear remarkably parallel to our own. But as we piece together the patterns of the past, we find that their world was much larger than we have imagined, a world in some respects so different from our own as to appear almost magical, a world so cosmic and a vision of man so fantastic that even the truth may seem like fiction—as it did in Marco Polo's day, 700 years ago.

CHAPTER TWO

# A 4000 YEAR OLD PICTURE

## Modern Interpretations of Myths and Legends Reveal a Sophisticated Science of the Stars

Myths are usually regarded as tales conjured up by early man to explain natural phenomena he did not understand. In the last few years, however, scholars have found knowledge recorded in mythic forms that further recasts our conception of the ancients.

In *At the Edge of History* William Irwin Thompson points out that myth is not an early level of human development. The most 'primitive' peoples investigated, Thompson says, are not the beginning of something but the end of something else. In *Hamlet's Mill* historians of science, Giorgio de Santillana and Hertha von Dechend, remark that the literature of the earliest peoples on record is concerned with terms and traditions that even for them were already 'tottering with age'. 'Those first predecessors of ours', they note, 'instead of indulging their whims with childlike freedom, behave like worried and doubting commentators: they always try an exegesis of a dimly understood tradition.'[1] Their major revelation is that many ancient myths have a common origin in a cosmology not terrestrial but celestial. What we have called 'myth' was in many cases actually a preliterate form of science; the gods and places of myth were but metaphors or ciphers for celestial activity.

One of the clearest examples of this change in dimension has emerged since the publication of *Hamlet's Mill*. Sixteenth and seventeenth century

*The* Carta Marina *of Olaus Magnus, a 16th century map showing sea monsters and a whirlpool called* Horrenda Caribdis *off the coast of Norway.*

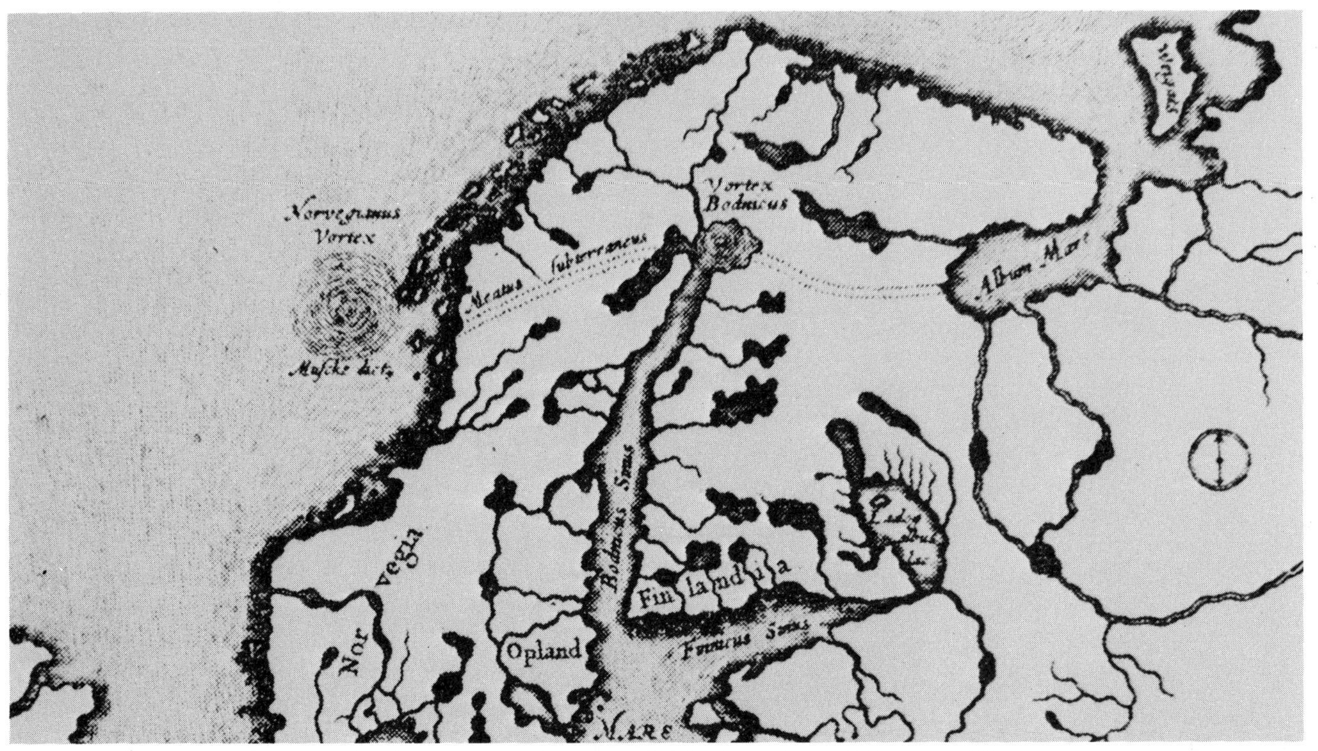


*A 17th century map showing the same whirlpool called the Maelstrom off the coast of Norway.*

maps of northwest Europe and adjacent waters show a great whirlpool, sometimes called the Maelstrom, off the coast of Norway. There is no great whirlpool in the sea off the coast of Norway today, and at first glance it may seem that this whirlpool is just another mythical terror in what was then a largely unexplored ocean. But this part of the ocean had been travelled for centuries. Perhaps there is another explanation.

Extensive research by contemporary scientists has shown that over great periods of time the positions of the earth's north and south poles have changed. By studying the patterns of magnetic alignments in rocks which permanently recorded the earth's magnetic field and other geologic and biologic records, researchers have been able to pinpoint the areas where the poles were once located. Since the poles have shifted more than once, evidence has emerged showing several former polar positions. The Chinese investigator, Ting Ying Ma, formerly with the University of Fukien, has concluded that the poles were once located near what is now Hawaii and Madagascar. Charles Hapgood has determined that possibly as early as 50,000 BC the north pole was located off the coast of Norway. In fact, the polar position determined by Hapgood neatly matches the location of the whirlpool shown on the old maps.

This may seem entirely coincidental until it is remembered that there *was* and *is* a great whirlpool in the sky, the centre of which precisely marks the celestial pole. The swirling motion of the circumpolar stars around the celestial pole is, from most latitudes, the most dramatic motion in the night

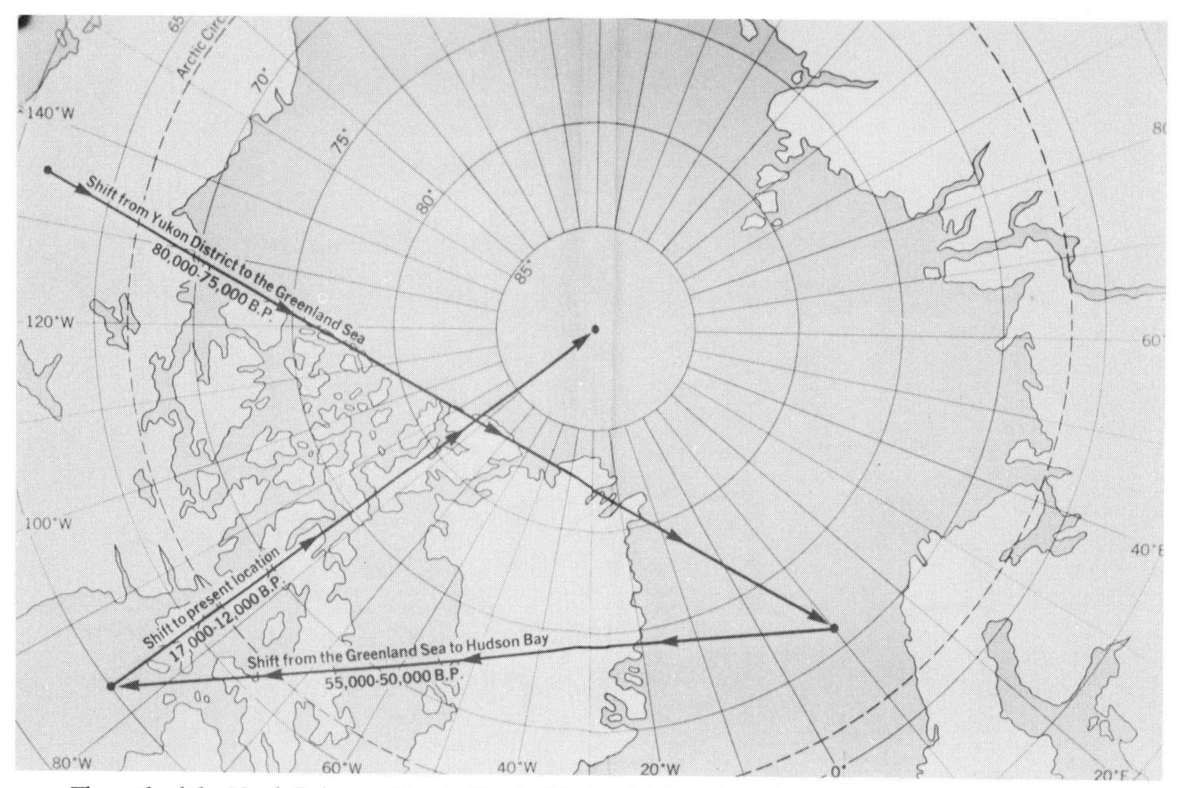


*The path of the North Pole according to Charles Hapgood. The pole position off Norway coincides with the location of the Maelstrom shown on early maps of the area.*

*The motion of the Circumpolar Stars around the celestial North Pole.*

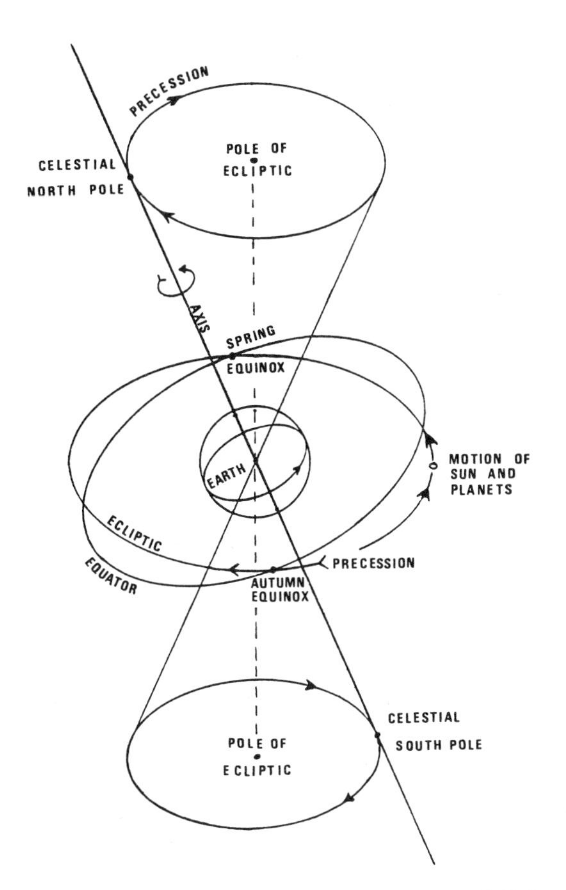


*The precession of the equinoxes. The equatorial North Pole is the celestial North Pole as seen from earth. This pole is determined by the earth's axis and revolves around the ecliptical North Pole over a period of approximately 26,000 years.*

sky. This motion, of course, is a result of the earth rotating on its axis; the celestial pole is *determined* by the direction in which the earth's axis points.

It would have been entirely natural, then, for an ancient cartographer to indicate the pole with a whirlpool, mirroring the whirlpool in the heavens. It seems that just as the outlines of Antarctica were passed on by unknowing scribes down to the 16th century, others again and again faithfully recorded a detail, the true meaning of which had been forgotten. What was long seen as a mythical feature of the earth becomes a celestial description from an old cosmology—a perfect example of how 'the gods and places of myth were but metaphors for celestial activity'.

A similar reinterpretation of a 4000 year old picture takes us much further. An Egyptian relief from about 2000 BC shows the falcon-headed god Horus and his opposite, the god Set, holding ropes intertwined around a pole. Above the pole are hieroglyphs enclosed with an oval forming a *cartouche*, a 'royal name'. The *cartouche* is translated 'Senusert', thought to be Senusert I, the first or second pharaoh of Egypt's Middle Kingdom (2040–1786 BC). The traditional interpretation of this relief is that it symbolizes the uniting of the two countries under Senusert. This unification of upper (southern) and lower (northern) Egypt is thought to have

*Horus and Set on a monument of Senusert I, about 2000 BC.*

been accomplished after the country had passed through an era of chaos of unknown duration.

There is much to recommend the traditional interpretation. Order and chaos run like alternating rhythms from historic times into the prehistoric horizon. The country was united at the beginning of the New Kingdom (1576 BC), the Middle Kingdom, the Old Kingdom and there are indications of a unification in predynastic times.[2] The theme of the unification of the two countries was an ancient one even for the pharaonic Egyptians, and having both political and religious implications, it held considerable importance. In fact, there are many reliefs in Egypt attributed not only to Senusert but to other kings as well, which all show the two opposing gods, the ropes and the pole in very much the same style as illustrated. The

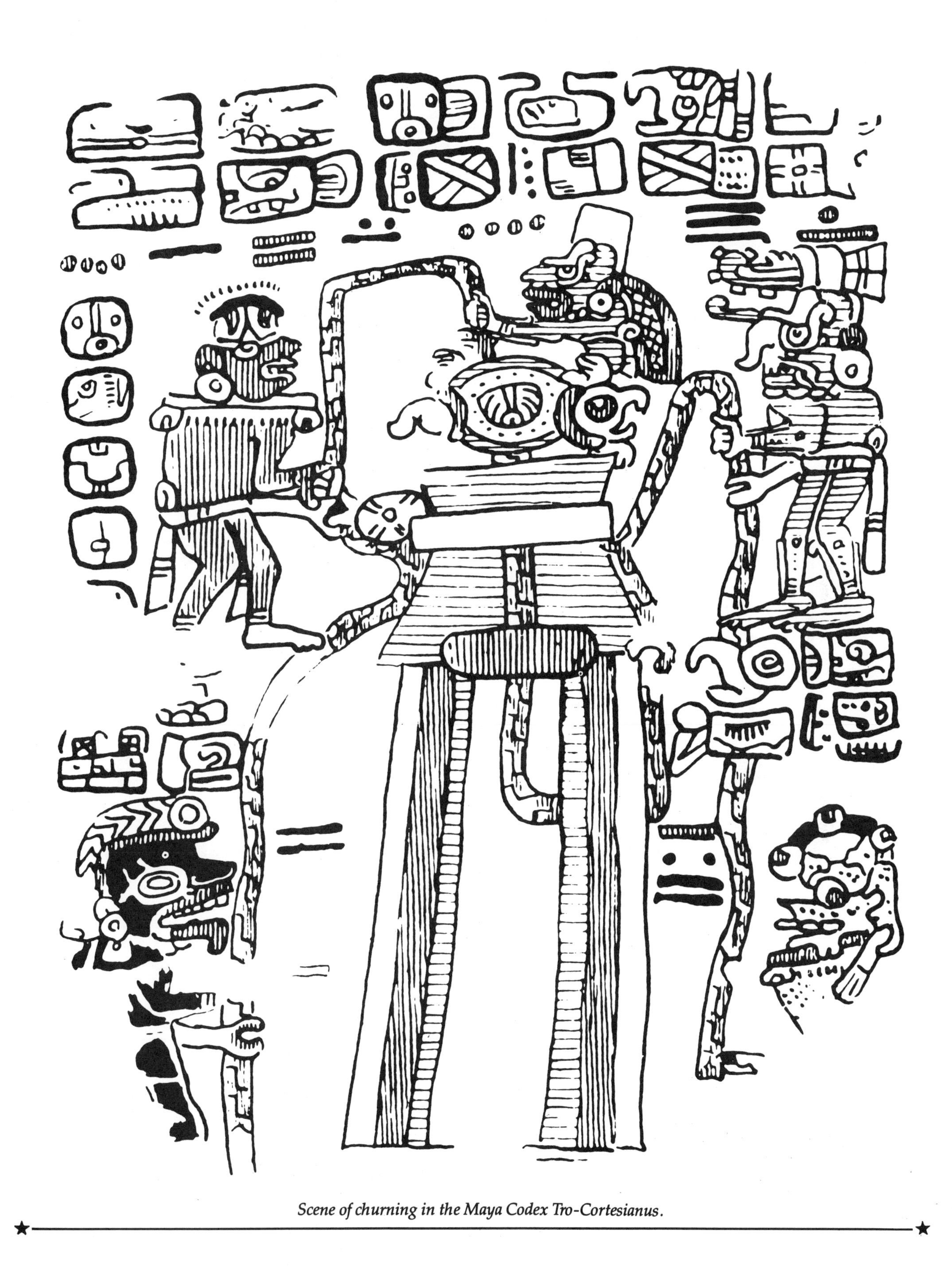

*Scene of churning in the Maya Codex Tro-Cortesianus.*

*A simplified version of the mighty churn that produced the constellations from the Milky Ocean.*

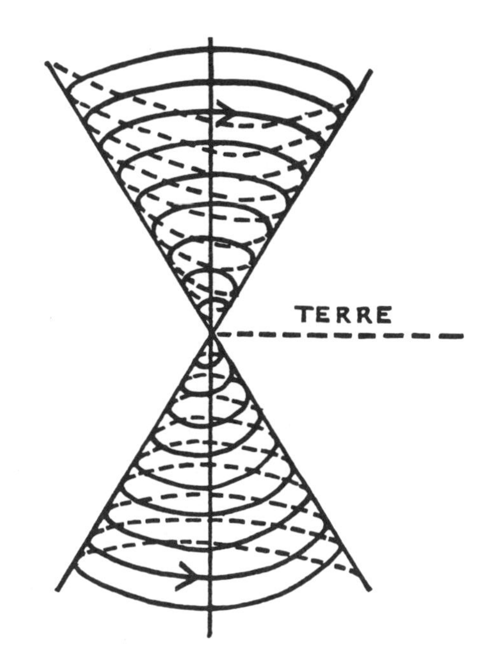


*The internal motion of the cosmic tree according to the Dogons of Mali, in northwest Africa, as recorded by a French ethnologist. This motion shows the rotation of stars above and below the earth around the fixed poles and the axis, which is otherwise symbolized by the cosmic tree.*

*The mighty churn of the Sea of Milk (the heavens) as described in the* Mahabharata *and* Ramayana.

interpretation of all these images is that they uniformly represent the uniting of the two countries. This seems about as certain as anything in Egyptology, since the uniting of the two countries was then a central function of the kingship.

However, with the significance of the whirlpool in mind, another interpretation of the relief is possible. Referring to this Egyptian relief in the context of similar images from cultures all over the world, de Santillana and von Dechend, fully aware of the standard interpretation, state that Horus and Set are not uniting the two countries: they are in the act of drilling or churning. And this churn is the cosmic churn that produces the great whirlpool in the sky.

With carefully detailed, massive research, the authors of *Hamlet's Mill* show that in ancient art, folklore, legend and myth the symbols of mills, churns, drills and whirlpools are all related. They are all metaphors suggesting turning and rotation. From the Dogons of Mali to the peoples of India and the Americas ancient sources repeat the same basic concept and image in which the milling, drilling and churning of the gods and heroes describes the motion of the circumpolar stars and the rotation of the 'world axis', an imaginary shaft which links heaven and earth. Supported by examples of art from three continents and by far more numerous references from folk tales and legends, de Santillana and von Dechend are as confident that our 4000 year old picture symbolizes the world axis as Egyptologists are that it refers to the uniting of the two countries.

From each perspective the apparently conflicting interpretations of Senusert's monument seem equally firm. It may not be a case, however, of having to reject one or the other. Many ancient Egyptian symbols have literally layer upon layer of meaning. Horus and Set may indeed be turning the cosmic churn, but this does not necessarily negate the traditional view that this monument celebrates the union of the two countries. Horus and Set were opposites: Horus, an alternate for Ra, was god of the sky by day and Set, also known as Typhon by the Greeks, was god of the sky by night.[3] In later times Set was considered the personification of sin and evil, just as Horus represented light and good. The union of these two gods in a common task was symbolic of union in general.

There is little doubt that de Santillana and von Dechend are also correct. Since the publication of *Hamlet's Mill* their astronomic interpretation has received further confirmation from the unlikely quarter of North American archaeology. In *America B.C.* Harvard professor Barry Fell documents the discoveries of Old World writings dating from before Christ in the Americas. Inscribed stones recording Celtic Ogham, Iberian, Punic, Basque, Phoenician and languages of North Africa indicate prehistoric transatlantic travel and commerce. Fell's thoroughness and abundance of information are such that the case cannot be dismissed.

This in itself is enough to revolutionize American archaeology, but Fell's most interesting find is an American Rosetta stone called the Davenport Stele, originally found in the lower levels of an Indian burial mound in Iowa in 1874. This stele, like many other American archaeological anomalies, was long considered fraudulent because investigators did not have sufficient training to identify, much less translate, the languages and scripts contained on it.

Fell had the background and knowledge to perceive that the stele contains inscriptions in three languages: hieratic Egyptian at the top, Iberian Punic along the upper arc and Libyan along the lower arc. Fell remarks that this stele is certainly genuine because neither the Libyan nor the Iberian scripts had been deciphered at the time it was discovered. He has translated the Egyptian text as:

> To a pillar attach a mirror so that at New Year the sun being in conjunction with the Ram in its house at the tilting of the balance in the Spring, the festival of celebration of the First of the Year and religious rites of the New Year are to take place when the Watcher Stone at sunrise is illuminated.

The Iberian Punic reads (from right to left):

> Set out around this is a secret text defining the season's delimiting.

The Libyan reads (from right to left):

> This stone is inscribed with a record. . . . It reveals the naming, the length, the placing of the seasons.[4]

Below the texts is a scene depicting the Egyptian celebration of the New Year on the morning of the vernal equinox, which now occurs around 21-22 March. Pulling on ropes, the celebrants are erecting a special pillar called the Djed, in this case a bundle of reeds surmounted by four or five rings. From an Egyptian tomb inscription of the 18th dynasty, Egyptologist Adolf Erman concluded that the ancient Egyptians annually set up the Djed column each spring in honour of Osiris.

## Is there Evidence of a Highly Developed Sense of the Order of the Universe Among the Ancients?

In Egypt the Djed column is a common hieroglyph found in texts from the earliest to the latest times.[5] The Djed column is often translated as symbolizing 'stability'. (It also has associations with vertebrae, the spine or backbone, and is sometimes referred to as 'the backbone of Osiris'.)

*Setting up the Djed. From a relief at Abydos.*

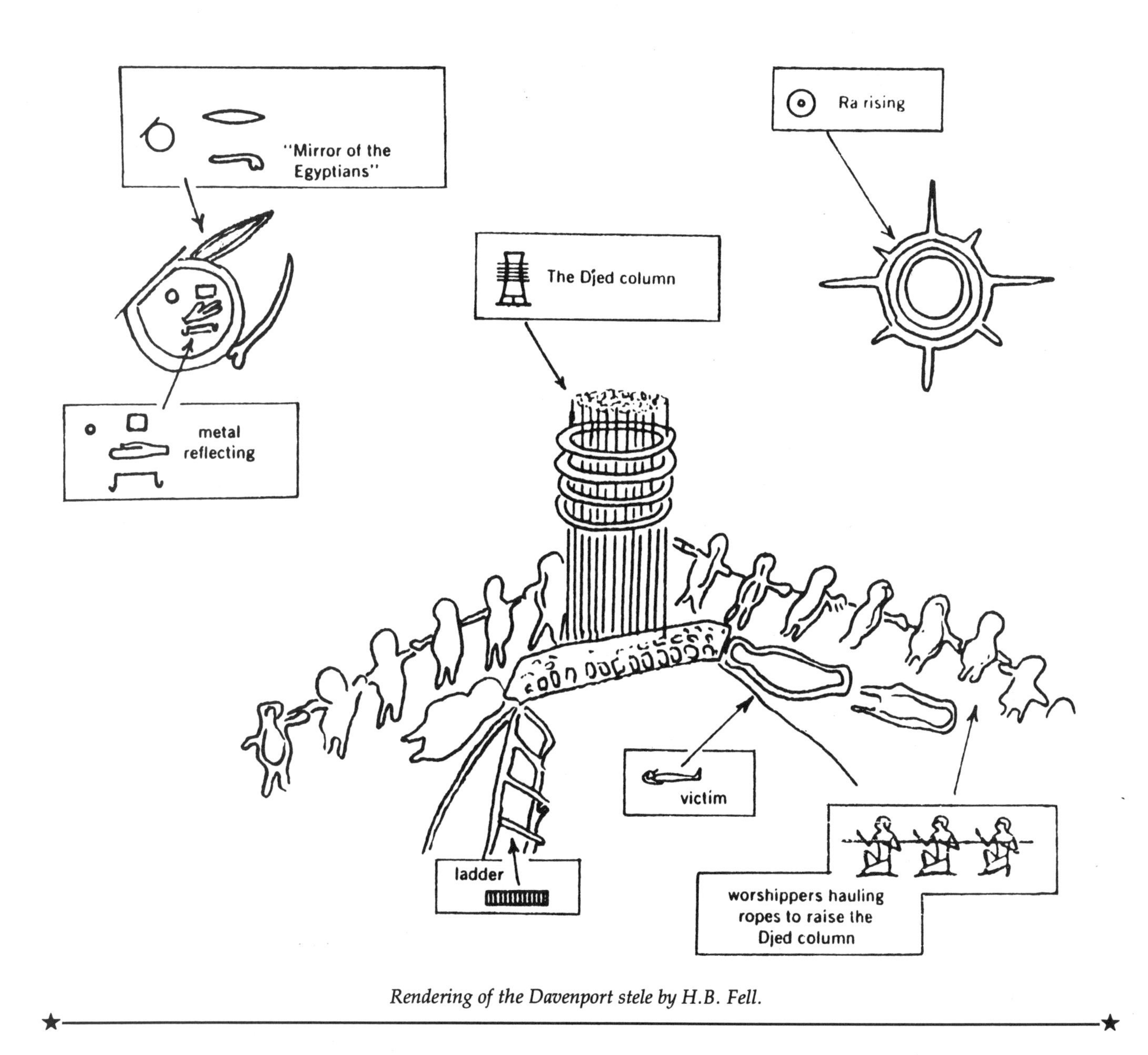


*Rendering of the Davenport stele by H.B. Fell.*

The Djed column also occurs in the inscription on Senusert's monument. That inscription reads to the effect that someone or something 'gives life, stability', though it is not clear whether it is Senusert or the rotation of the axis by Horus and Set that produces these effects. Of course, it is the daily rotation of the earth upon its axis that brings day and night, the tilt of the axis with respect to the plane of the earth's annual revolution around the sun that results in the turning of the seasons, and the slow wobble of the axis that produces the precession of the equinoxes against the background of fixed stars (and around the ecliptical poles) over a cycle of nearly 26,000 years. The fact that the earth turns, tilts and wobbles upon its axis is the essential key in accounting for the life-sustaining, stable motions of the heavenly bodies. If the earth did not rotate, the sun would not rise and set and there would be little or no life on our planet.

Ancient peoples are not supposed to have known that the earth rotates upon its axis, and that it is this rotation of the earth that accounts for the circular motions of the stars around the celestial poles. According to most theories, ancient peoples believed that the earth was flat and that the sun, planets and stars move around it like the moon.

De Santillana and von Dechend stop short of claiming that ancient peoples believed in 'the abstract idea of a simple earth axis, so natural today'. As they perceive ancient cosmology, the ancients 'thought of the whole machinery of heaven moving around the earth, stable at the centre.'[6] They explain that in myth one line always implied many others in a structure. While ancient images relating to mills, poles, axles and trees clearly represent the world axis, this *axis mundi* is itself an abbreviation, more or less, for the entire moving frame of the cosmos to which it is connected.

In the case of the Djed column, however, (the symbology of which de Santillana and von Dechend do not explore), the Egyptians came remarkably close to symbolizing the abstract idea of a simple earth axis. Some would say they obviously achieved it. The Egyptians represented their cosmic and religious beliefs with symbols modelled after material forms in order to assist in the expression of abstract concepts. Cities and places in higher realms were fashioned after counterparts on earth; a man's soul was depicted as a bird; and the symbol of life-giving energy was the *ankh*. There are many sources suggesting that this symbolic stylized column reflects the fact of the earth's rotation and that it actually represents the abstract idea of a simple earth axis. The transparent symbolism of the Djed as a pillar which even looks like it is spinning testifies that their concept of an axis was very similar to ours.

Further evidence of the rotational symbolism of the Djed may be seen in vignettes from ancient Egyptian papyri. In two remarkable examples depicting the sunrise, the Djed directly supports the *ankh*, the symbol of life-sustaining energy, which in turn supports the rising sun which sails beneath the vault of heaven, clear examples of how the Egyptians used obvious symbols to communicate abstract ideas. In another illustration Horus-Ra, the sun god, wears a disc encircled by a serpent and again stands beneath the vault of heaven. As in the other cases, his position 'rises from' a Djed column beneath him, which has human arms and hands holding symbols of power.

Other indications that the ancients had an understanding of the earth axis as well as a world axis come from Greece. In the second century after Christ, the Greek astronomer Ptolemy put the earth at the centre of the solar system, which was the position it retained in most quarters until the

*A mummy lies upon a bier; above is its soul in the form of a human-headed bird holding in its claws the emblem of eternity,* shen.

time of Copernicus, 14 centuries later. However, before Ptolemy the Greeks had a heliocentric theory of the solar system and lost it. In the first or second century before Christ, a Greek named Aristarchus put forward a heliocentric theory. Around the same time a countryman named Heraclides explained the variations in the brightness of Venus and Mercury by maintaining that they revolve around the sun, not the earth. Heraclides also explained that the apparent motions of the heavenly bodies across the sky were due to the earth spinning on its axis.[7]

Given the universal symbolism of the world axis and the many indications that scientific conceptions and precision were *in decline* during historical antiquity, earlier Egyptians and other ancient peoples could well have made the same deductions. The Nesi-Khonsu papyrus of about 1000 BC refers to 'the circuit of the earth' and 'the earth in its courses',[8] two suggestions that the Egyptians, too, knew that the earth travels around the sun. With skeletal remains showing that our species has been on earth for 100,000 years, it is hardly likely that even the Egyptians were the first to develop these ideas.

We can see today that ancient peoples had far more developed concepts of the order of the universe than has long been thought. And suddenly we are in a different world: the past has changed. Not only do we find that

*An ancient plaque with the* was *sceptre, a Djed and an* ankh. *The* ankh *symbolized life and the forces of life; the* was *sceptre symbolized the power to destroy.*

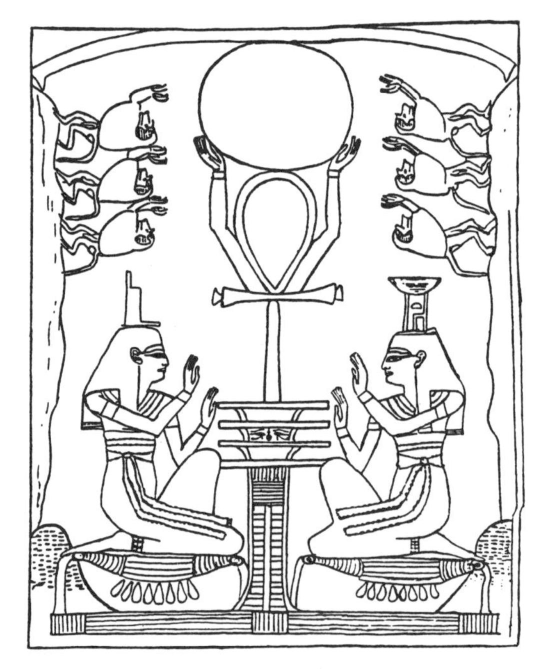

*A depiction of the sunrise from the Papyrus of Ani, about 1000 BC. A Djed column which holds the body of Osiris stands between Isis and Nephthys. The Djed supports an* ankh, *the emblem of life, which has arms supporting the disc of the sun. Above the sun is a symbol for the vault of heaven. The six apes represent the spirits of the dawn.*

*A Djed sunrise: the Djed supports an* ankh, *symbol of life-giving energy, which in turn produces the rising sun. Here too the six apes are the spirits of the dawn. From the Papyrus of Qenna, about 1000 BC.*

*Horus-Ra, wearing a sun-disc encircled by a serpent, stands beneath the vault of heaven. He is worshipped by seven apes, spirits of the dawn, and stands above a Djed with arms and hands holding symbols of the power of Osiris. The Djed is flanked by Isis and Nephthys. From the Papyrus of Hu-nefer, about 1000 BC.*

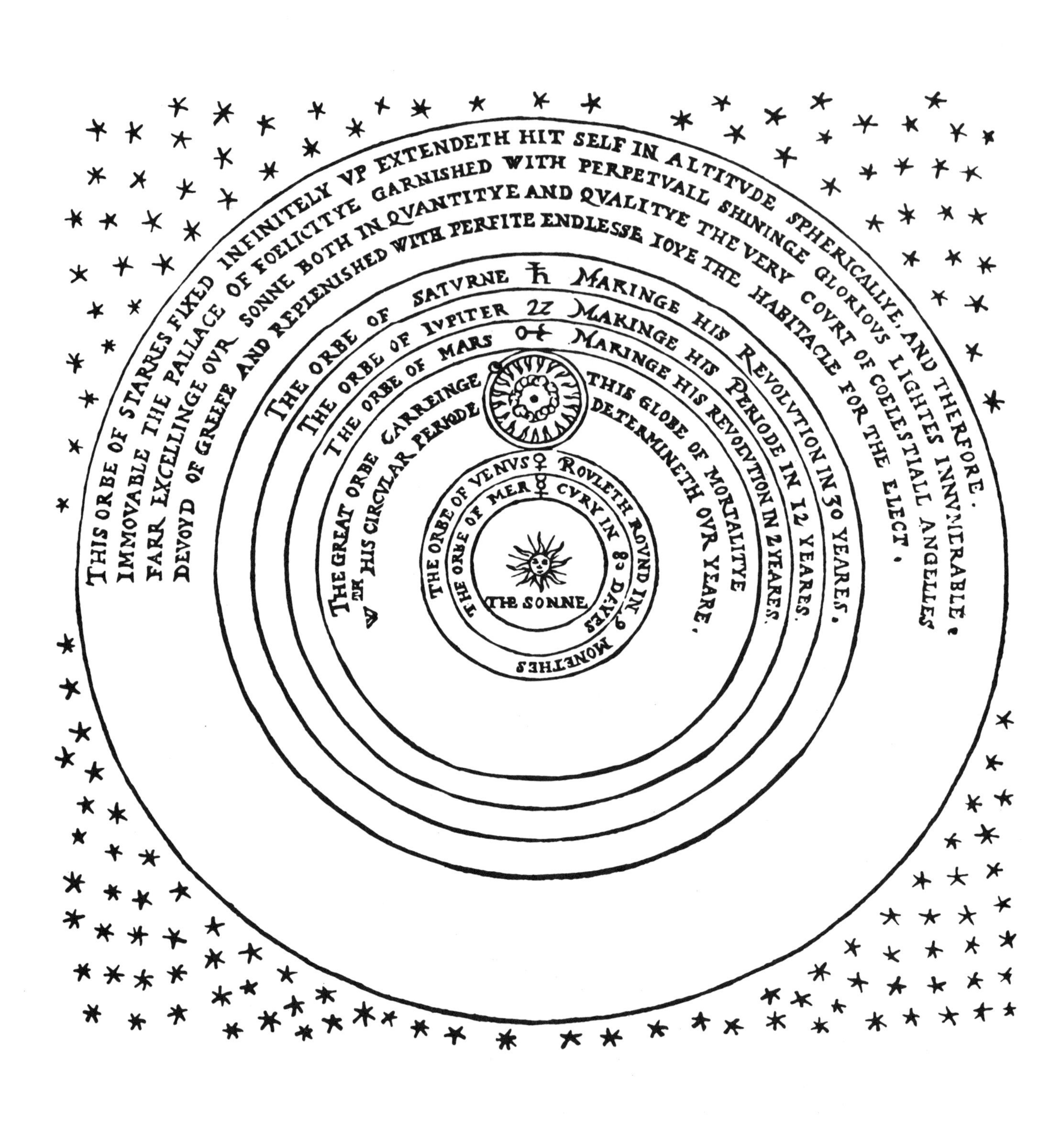


*The Copernican conception of the solar system with the sun at the centre and the moon orbiting the earth.*

*The spirits of Ra (as Horus) and Osiris flanked by Djeds.*

Egyptian calendar feasts took place in the Americas 2000 years before Columbus, but we discover ancient symbolism reflecting the earth's rotation that is remarkably consistent from remains of three continents.

The most interesting aspect of this new world is the way the old symbols lead on to new and far-reaching implications. The Egyptians regarded the Djed and its meaning with a great deal of importance. It is one thing to find the Djed involved in depictions of the sunrise, symbolizing an axis, but it also occurs frequently in vignettes concerned with travels of the soul. Small stone images of the Djed are found as amulets interred with burials dating back to predynastic times.

If we remember that the earth's axis defines the celestial pole and the centre of the circumpolar stars, one suspects that the Djed's significance connected in some way with their entire conception of the cosmos. The strongest and most important indication that this was so is found first of all in the Great Pyramid and, indeed, in all the major pyramids of Egypt.

*Small stone Djed amulets found in tombs by Flinders Petrie.*

*A scene from the Turin Papyrus showing a spirit of the dead travelling in a boat like the bodies of heaven. Also in the boat are a symbol of Ra (the falcon's head) and a Bennu bird, an Egyptian equivalent of the phoenix, a symbol of immortality and the birth of the spiritual body from the physical body. In front of the boat is a table of offerings, Osiris, god of the dead, and a Djed column.*

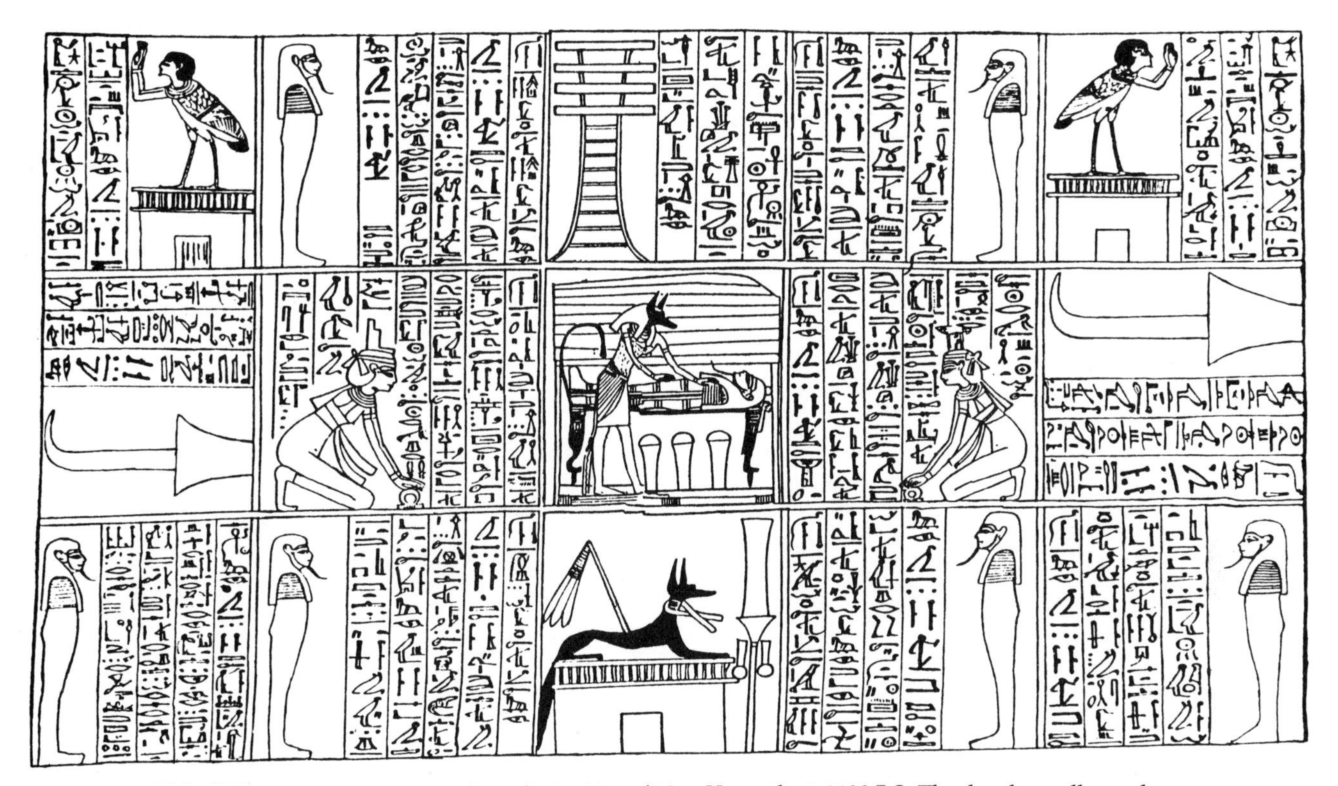

*A depiction of a mummy chamber from the papyrus of Mut-Hetep about 1100 BC. The chamber walls are shown as lying flat. In the centre Anubis, god of the dead, presides over the mummy. Above him is a Djed column, presumably on the ceiling, indicating the upward path. Below, Anubis in the form of a jackal rests upon the entrance to the underworld. Left and right of the central panel are the goddesses Isis and Nephthys. The four inner bearded figures are gods symbolizing the four cardinal points; the two outer figures on the bottom may represent the zenith and nadir, or may be* shabtis. *In the upper right and the upper left are souls in the form of birds; in the panels below them are votive flames.*

CHAPTER THREE

# STAR TEMPLES

## Probing the Mystery of the Pyramids

The pyramids rise like spectres from another age, challenging comprehension. It may be that some pyramids record basic geometric and mathematical data and that at a certain ratio the external plan of the Great Pyramid encodes the key dimensions of the earth. Even if preservation of knowledge was a key function of some pyramids, this purpose cannot be said to explain them all, nor does it entirely account for even those which do reflect scientific knowledge. Inside all the pyramids there are chambers and galleries whose primary function cannot be explained in terms of recording scientific concepts. What took place *inside* the pyramids?

Naturally, as ancient and famous as these buildings are, many opinions abound as to what took place inside these inner chambers. Among Egyptologists the opinion is almost universal that these chambers held the bodies of dead kings. The strongest suggestion of this is that the chambers in some pyramids have large stone boxes called sarcophagi, usually with lids, and just the right size to hold a coffin, a mummy case or a man.

If one does not question why tombs should encode scientific information or why they should be so large, the belief that the pyramids were tombs at first glance seems natural enough. But a belief it is, and not knowledge. There is a very great difference between what is believed about the past and what is known, and this is especially true of the Egyptian pyramids. What *is* known is that there has never been an original burial found in any pyramid in Egypt.[1] Grave robbery has long been and still is

*The pyramids of Gizeh.*

*The moon over the Second Pyramid.*

the standard explanation for all the missing bodies and associated burial artifacts which the tomb theory postulates.

The belief in the tomb theory is still vehemently defended despite increasing indications of its inadequacy and some direct evidence against it. For instance, the chambers in some pyramids are not even large enough to hold a sarcophagus. In 1952 the excavation of buried remains of a much ruined pyramid at Saqqara (now called the Pyramid of Sekhemket) revealed a sarcophagus inside chambers which had been blocked up and undisturbed since remote antiquity. This appeared to be the first intact, original burial ever found in an Egyptian pyramid. The lid of this sarcophagus was sealed in place with cement, yet when it was opened it was perfectly empty. The Pyramid of Sekhemket contained no body or significant burial artifacts and yet had not been robbed.[2]

This is not the only example of such a find. Another sealed empty sarcophagus was found in 1925 by George Reisner and Wm. S. Smith in a chamber at the bottom of a 30 metre (99 foot) deep shaft about 90 metres (100 yards) east of the Great Pyramid. This is the only undisturbed 'burial' attributed to the Old Kingdom found to date in Egypt. In *The Riddle of the Pyramids*, Kurt Mendelssohn reports that another empty sarcophagus, also undisturbed for thousands of years, was found in a gallery under the Step Pyramid at Saqqara.

In fact, then, not only is there no clear and solid evidence of any kind that the pyramids were originally intended as tombs, there is direct evidence against the tomb theory.[3] But if they were not tombs, we still need an explanation of the chambers and galleries inside the pyramids.

One of the key patterns to be noticed in the pyramids is that the main passages in almost all of them point toward the circumpolar stars around the celestial north pole. In the Great Pyramid, the Descending Passage points almost exactly to the celestial pole, missing it by only some three degrees.[4] This does not necessarily mean that the Pyramid builders attempted to align the passage with the pole but missed by that margin. The extreme accuracy of the orientation of the Pyramid's sides to the four cardinal points of the compass, accurate to within a few minutes, indicates that its builders could probably have aligned the passage with the pole perfectly, if they had so desired.

It is far more likely that the passage was aligned with what was then the Pole Star. The definition of a pole star is not that the star must be exactly at the pole; it is simply a bright star, easily seen from earth, which is *closest* to the pole at any particular time. Because the earth wobbles slowly on its axis over a period of about 26,000 years, the position of the celestial pole

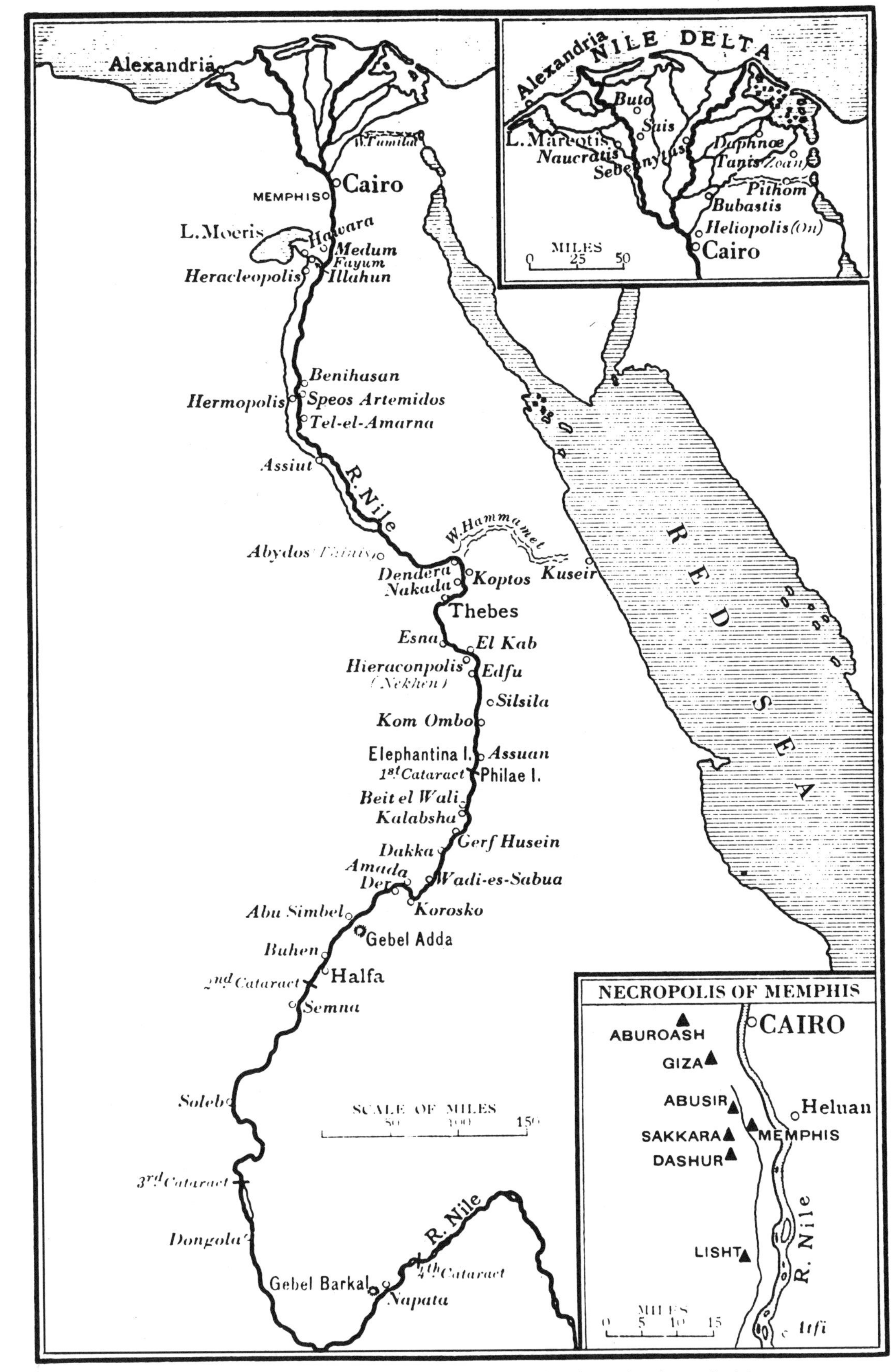


*A map of ancient Egypt and Nubia, showing the locations of major antiquities.*

constantly changes, gradually marking out a great circle in the sky. Over thousands of years, different bright stars become the pole star as the pole moves through the heavens. Today the pole star is Polaris, about one degree from the pole.

Several investigators have noted that the star Thuban (alpha Draconis) in the tail of the constellation Draco, the Dragon, would have been visible from the Descending Passage in 3440 BC and again in 2160 BC. But saddled with a conventional construction date for the Great Pyramid of 2700 or 2600 BC, most investigators have found little satisfaction in alignments to Thuban.

To which star did the passage point? Despite the currently widespread acceptance of the 2600 BC date, there is actually very little evidence to support this dating. Recent discoveries detailed in *Pyramid Odyssey* suggest a much more ancient construction date for the Great Pyramid—10,500 BC.[5] At that time either Rastaben or more likely Eltanin (beta and gamma Draconis), important stars sometimes known as the Eyes of the Dragon, may then have been visible from the Descending Passage.

It is extremely interesting that in *The Dawn of Astronomy* the British astronomer, Sir Norman Lockyer, noted that gamma Draconis or Eltanin was identified with the god Set[6]—the same Set who appears on Senusert's monument turning the pole with Horus. Set, as already mentioned, was god of the sky by night. But Lockyer—most of whose findings have not been assimilated by Egyptology—points out that the particular part of the sky with which Set was identified was the circumpolar stars.[7] Lockyer further notes that Eltanin was once the pole star and went on to identify one 'for certain' and possibly several Egyptian temples that were oriented to Eltanin after it had ceased to be the pole star and, from the latitude of Egypt, was no longer even circumpolar.[8] These clues suggest that the Egyptians may indeed have been watching the stars as early as 10,500 BC and strengthen the case for the Great Pyramid's alignment to Eltanin.

## The Pyramids as Celestial Observatories

In archaeology, when a temple or other ancient site has a significant alignment to a key rising or setting position of the sun or moon, the site is usually identified as a lunar or solar temple or observatory. Many of the stone circles of the British Isles, for example, are now described as 'megalithic lunar observatories'. From this perspective, described in terms of the alignment of its foremost passage, the Great Pyramid emerges as a

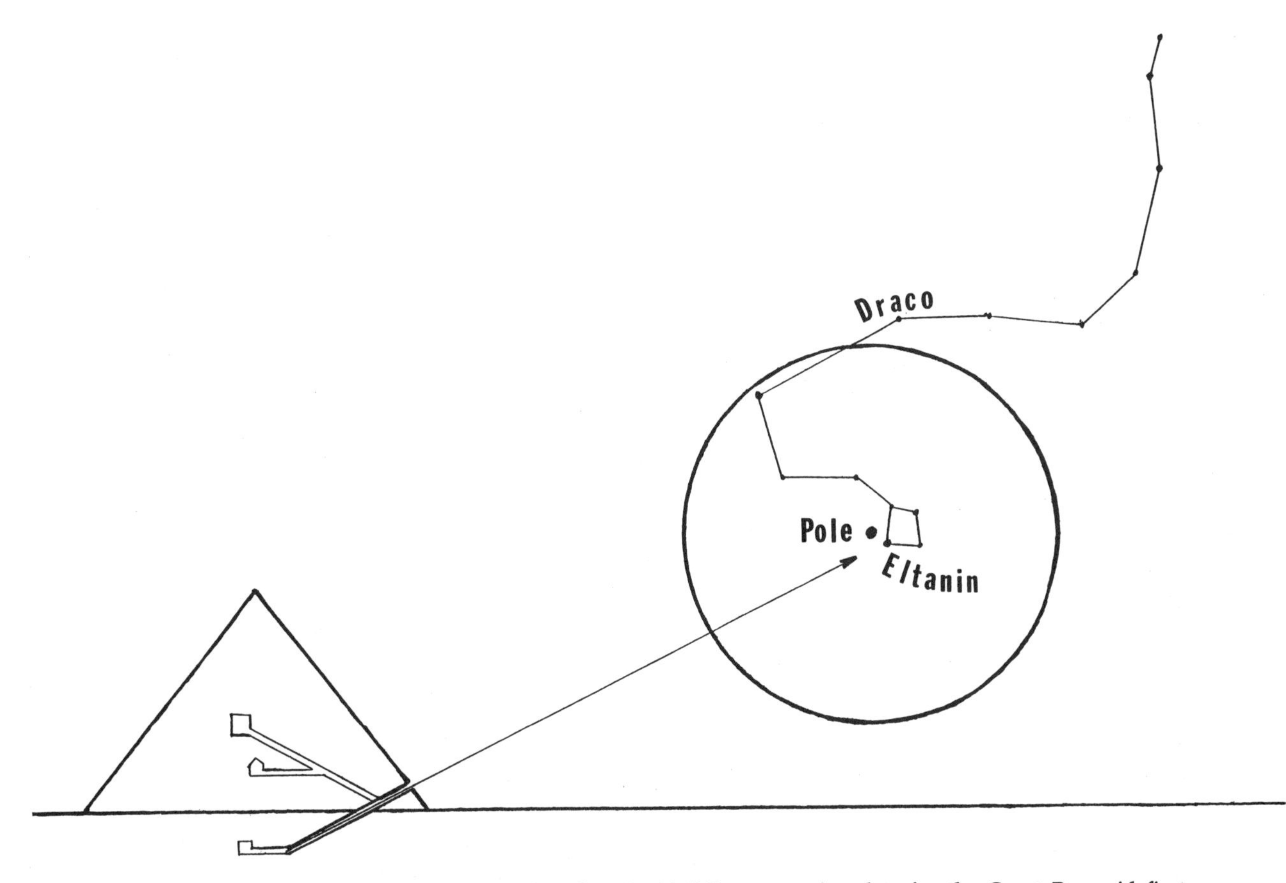


*Evidence has recently emerged supporting the 10,500 BC construction date for the Great Pyramid first advanced by the American psychic Edgar Cayce. At that time the Pole Star was Eltanin (gamma Draconis), one of the 'eyes' of the large constellation Draco (The Dragon). Eltanin was long regarded as an important star by the Egyptians. According to British astronomer, Sir Norman Lockyer, several Egyptian axial temples show evidence of alignments to Eltanin long after it had ceased to be the Pole Star.*

star temple. But this is true not only of the Great Pyramid; virtually all the pyramids of Egypt have their major passages aligned toward the north polar region. By the same criteria, they too are star temples.

The only major pyramid in all Egypt which does not have a shaft sighted toward the celestial pole is the Step Pyramid at Saqqara. Yet even this pyramid has a unique and dramatic feature which confirms the pattern: a small cube-shaped room or capsule built on to the north side of the pyramid. This capsule, called a *serdab*, has two peepholes through which can be seen a seated, life-size statue of a man said to be Zoser, the builder of the pyramid. *Serdab* statues such as this are thought to have been executed in the likeness of the king, so that his spirit would have a familiar form in which to rest. The unusual thing about this *serdab* of Zoser is that it is tilted back at an angle so that the statue faces eternally toward the celestial north pole. In all archaeology there is little which is more evocative of images from the space age. It almost looks as if Zoser were poised for flight.

The Egyptians' fascination with the stars was by no means restricted to alignments in pyramids and a temple or two. Lockyer has shown that

*The Step Pyramid of Zoser at Saqqara.*

*The stone capsule or* serdab *on the north side of the Step Pyramid at Saqqara. The capsule is tilted so that it is aimed toward the celestial pole.*

*The seated life-size statue of Zoser inside the* serdab *on the north face of the Step Pyramid. The statue faces the circumpolar stars.*

practically all of the great Egyptian temples (other than the pyramids) were aligned to serve astronomic functions. These alignments include orientations not only to many important stars but to risings and settings of the sun and moon. Although Lockyer's work has been largely disregarded in Egyptology, this has not happened for a lack of evidence on his part. Lockyer pointed out that many of the large temples meet all the requirements for 'horizontal telescopes'—instruments designed for the observation of bodies as they rise and set on the horizon. Probably the best example of this is the great Temple of Karnak which Lockyer described as 'beyond all question the most majestic ruin in the world'. It covers about twice the area of St Peter's in Rome and in total is nearly 548 metres (600 yards) long. Like most Egyptian temples it is organized around a central axis, which at Karnak is approximately 457 metres (500 yards) in length. As Lockyer put it: 'The whole object of the builder of the great temple ... was to preserve that axis absolutely open; and all the wonderful halls of columns and the like, as seen on one side or other of the axis, are merely details.'[9]

These details, however, also demonstrate the Egyptians' seriousness as astronomers. The axis passed through 17 or 18 narrow apertures in various pylons and walls, intercepted only by the sanctuary at the closed end of the temple. From better preserved temples elsewhere we can gather that the closed end was originally covered, making the sanctuary dim or dark. Lockyer pointed out that in modern telescopes, 'we have between the object-glass and the eyepiece a series of what are called diaphragms ... a series of rings right along the tube, the inner diameters of the rings being greatest close to the object-glass, and smallest close to the eyepiece'.[10] The purpose of these diaphragms is that the light from the object-glass shall fall upon the eyepiece without loss or reflection. The 17 or 18 apertures through which the Karnak axis penetrates become gradually smaller as they approach the sanctuary. Lockyer concluded: 'These apertures in the pylons and separating walls of Egyptian temples exactly represent the diaphragms in the modern telescope.'[11]

The long axis at Karnak pointed to the setting sun on the longest day of the year. With the system of apertures, a darkened sanctuary and the length of the axis, the target in the sanctuary would be illuminated by the rays of the setting sun for only a few moments on one particular day each year. Utilizing these features, the Egyptians could determine the length of the solar year with a precision of about one minute.

Karnak was a solar temple but its features as a horizontal telescope are found in many other Egyptian temples. The evidence is overwhelming that many of these temples were used to observe important stars. In the first

*The axis of the Karnak Temple, looking east.*

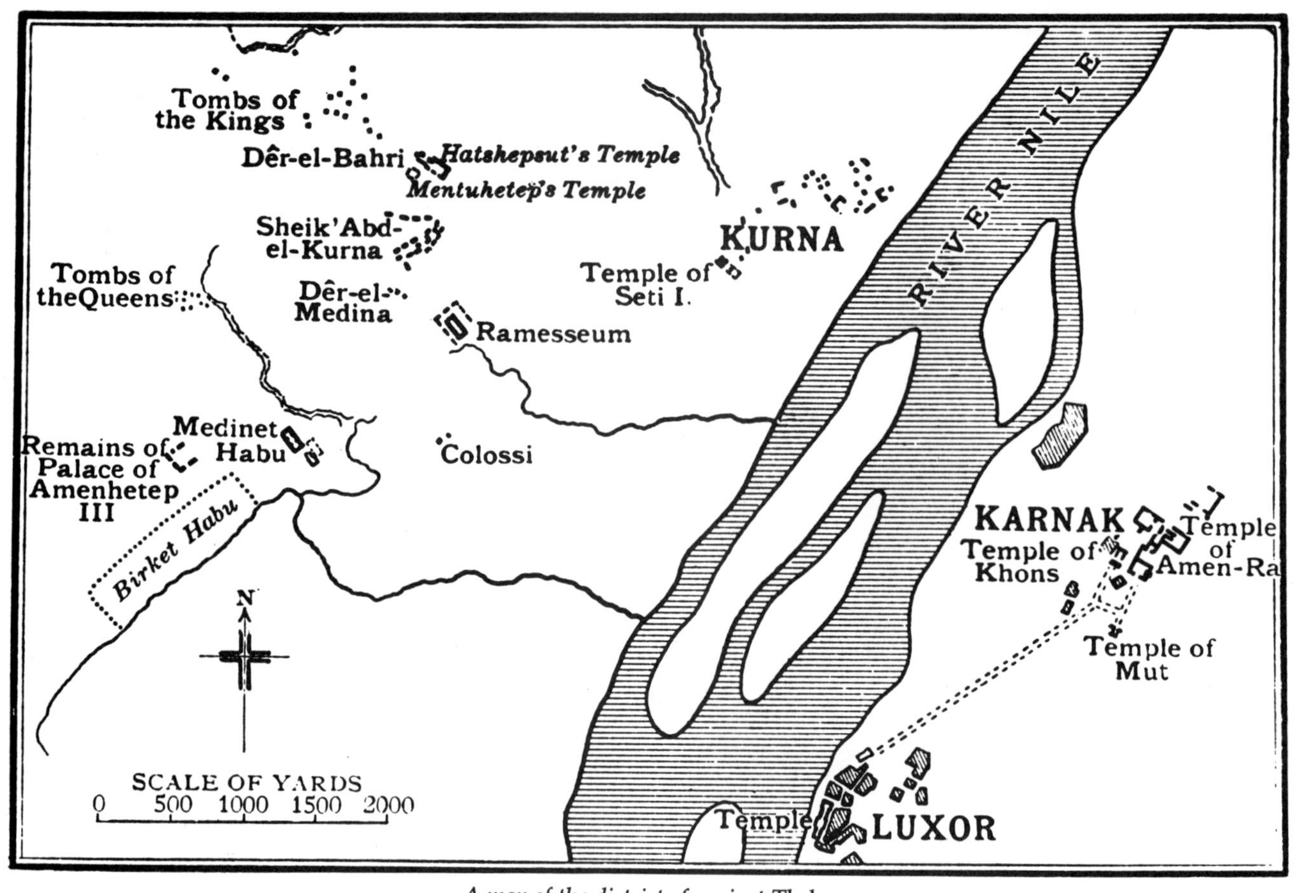


*A map of the district of ancient Thebes.*

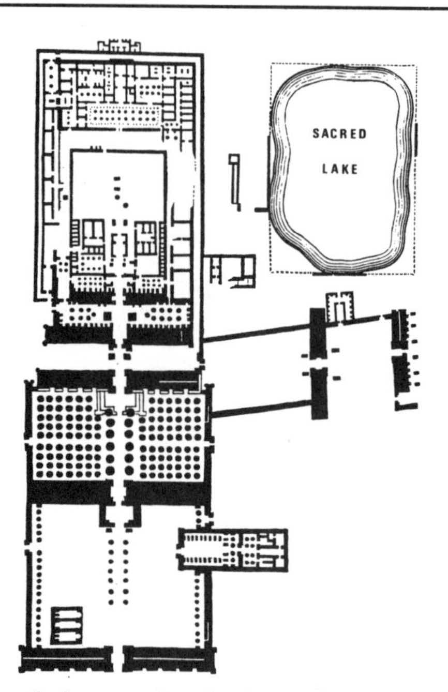


*Plan of the Karnak Temple showing the long temple axis orientated to the setting summer solstice sun. The plan also indicates an extra external gate located on the extended axis line. The Sacred Lake may have been used in some of the boating rituals of the Heb Sed festival. Smaller later temples, also astronomically orientated, were built on to the main complex.*

*The axis of the Karnak Temple as it appeared in the 19th century, before restoration.*

*A monolithic gate made of lava at Tiahuanaco, Bolivia, and strikingly similar to the Gate of the Sun on the Greek island of Naxos. The principles of astronomy as practised in Egypt and Greece were almost certainly employed by the ancient high cultures of Peru and Bolivia.*

*The Gate of the Sun, Tiahuanaco, Bolivia. This massive gate has often been remarked upon for the engineering and artistic prowess required to execute and position it. Like other gates at Tiahuanaco, in Greece and in Egypt, its chief purpose may have been to provide a firmly fixed aperture through which to make astronomic sightings.*

*The ruins of Karnak and the Sacred Lake circa 1880.*

place, ancient building inscriptions have survived describing a ceremony called 'stretching the cord' during which the foundation of the temple was dedicated and aligned with a star. The star used for the alignment is actually named in two cases. Regarding the alignment of the temple of Hathor at Denderah, the inscriptions state that while stretching the cord, the king directed his glance to the *ak* (probably alpha Ursae Majoris[12]) of the constellation of the Thigh, an Egyptian symbol for the constellation now known as the Great Bear. This ceremony took place around 300 BC when a new temple was built on the foundation site of a much more ancient structure which had been destroyed. The inscription states that the ceremony was a reenactment of an earlier one 'as took place there before'. The actual inscription has been translated as follows:

> The living god, the magnificent son of Asti, [a name of Thoth] nourished by the sublime goddess in the temple, the sovereign of the country, stretches the rope in joy. With his glance towards the *ak* of the Bull's Thigh constellation, he establishes the temple-house of the mistress of Denderah, as took place there before ... I establish the corners of the temple of Her Majesty.[13]

In temples where such inscriptions have not survived, their use as stellar observatories can still be deduced from their axes. Because of the gradual precessional motion of the earth's axis over 26,000 years, the rising and setting positions of the stars are slowly but constantly changing. With the narrow apertures penetrated by the long axes of the Egyptian temples, a star seen rising through the outer gate or propylon would no longer be visible 200 or 300 years later. Because of this precessional motion and the king's responsibility to maintain the calendar, the early kings were each

*A reconstructed view of the Hypostyle Hall on the central axis of the Temple of Karnak, which Sir Norman Lockyer described as 'a horizontal telescope'.*

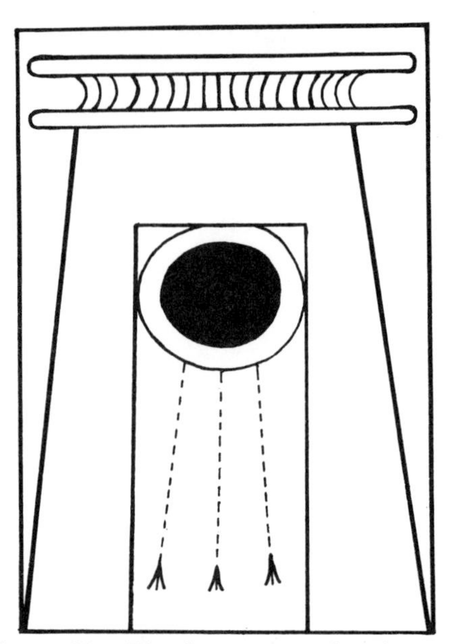

*Sun in gate, direct testimony from a New Kingdom papyrus that the Egyptians sighted the sun through astronomic gates. This vignette is a detail from Papyrus M in Alexandre Piankoff's collection,* The Wandering of the Soul. *Papyrus M is a text concerning the divinities and divisions of the Underworld.*

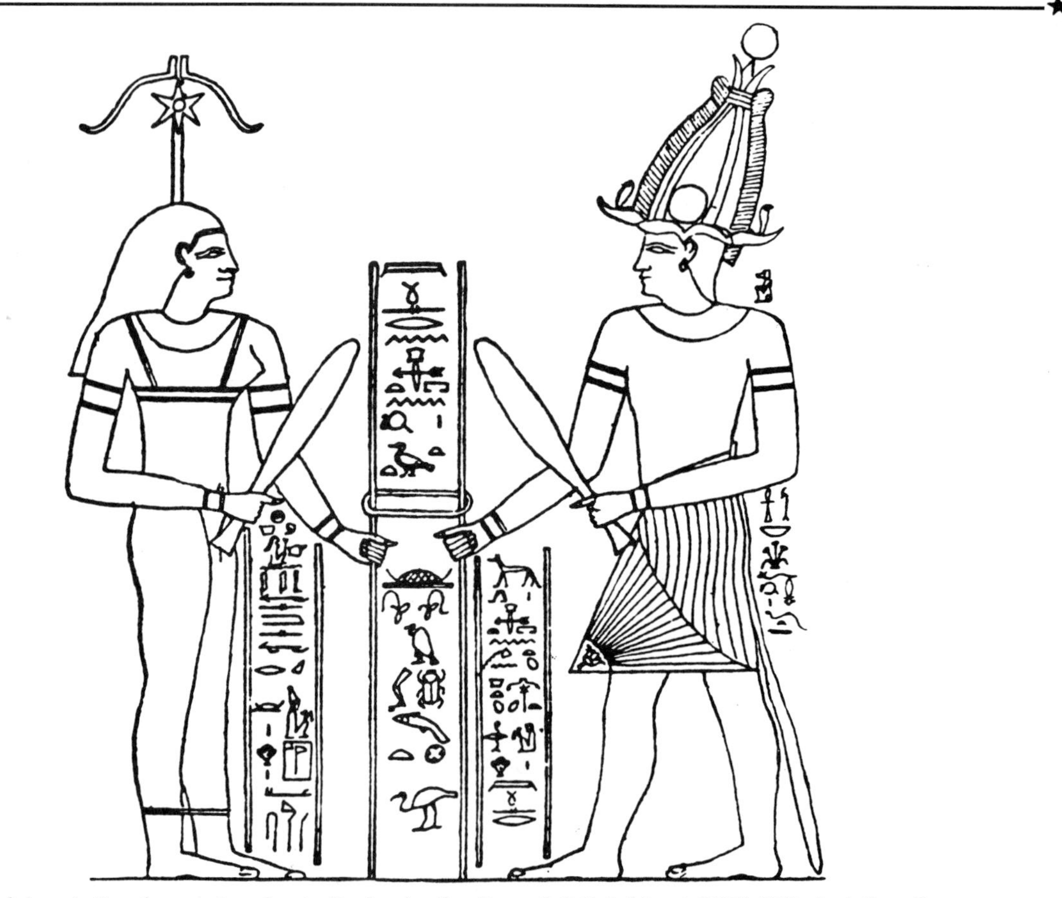

*A relief and inscription from a temple at Abydos in the time of Seti I (about 1300 BC) depicting the foundation-stone ceremony in which the temple was orientated to a particular star. The goddess Sesheta, with a star above her head, faces the king, who wears the crown of Osiris. Each holds a long peg and a club for driving the pegs into the ground. Around the pegs runs a taut cord. Note that the king's kilt includes a tail.*

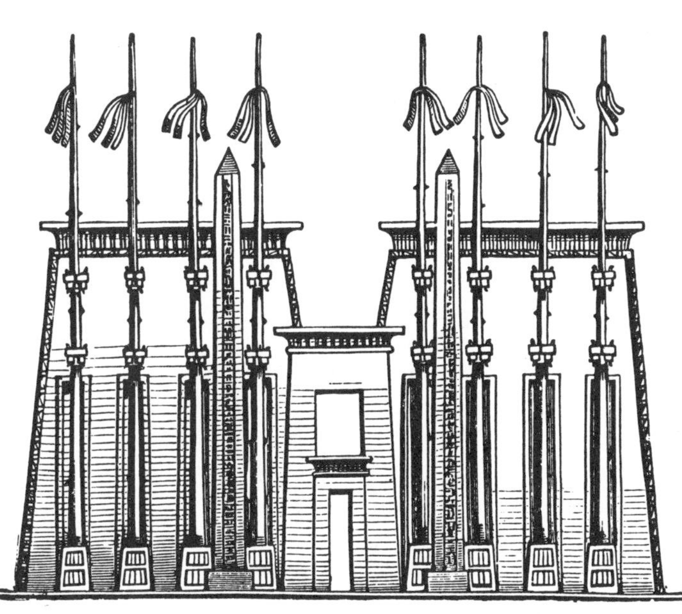

*A reconstruction of a typical Egyptian temple gate and outermost pylons showing the narrow opening of the entrance and a window above it. This window corresponds to astronomic windows at the New Grange mound in Ireland and the Treasury of Atreus at Mycenae in Greece.*

*The huge Gate of the Sun or Gate of Apollo on the Greek island of Naxos.*

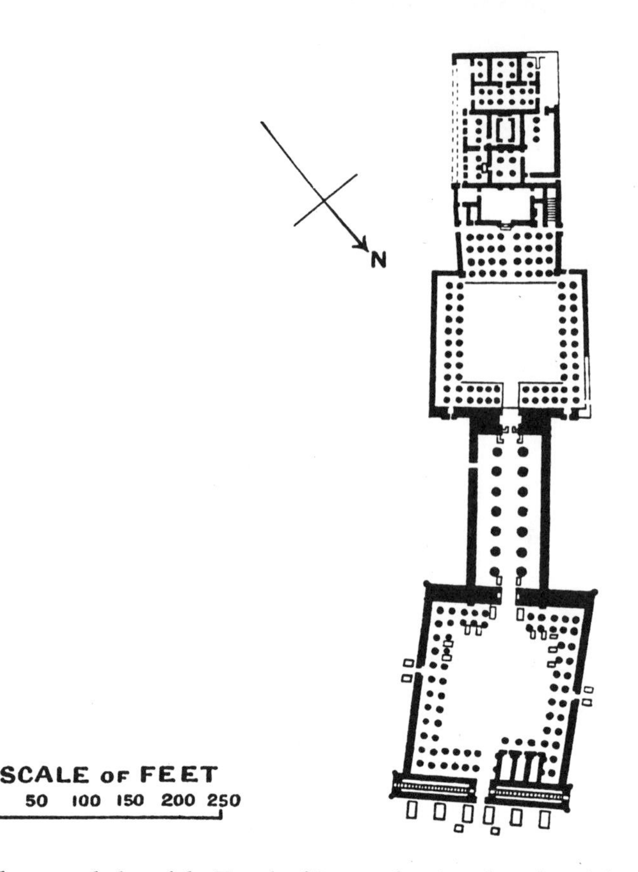


*The ground plan of the Temple of Luxor, showing alteration of the temple axis to accommodate the changing rising position of the star to which it was orientated.*

expected to build a new temple, a practice that was later discontinued due to its expense. If the Egyptians wished to continue sightings of a particular star, it would have been necessary either to alter the temple axis or to build another temple with a slightly different orientation. Such changes in the axis actually exist in the Luxor temple, where four separate additions with slightly different orientations were made end-on to the old axis. At Medinet Habu, across the Nile from Luxor, where there was ample space next to the old temple, another entire temple was built with an axis orientated a few degrees from the old alignment.

In still more cases, Lockyer found that slowly varying positions of important stars such as Eltanin, Phact, Canopus and Sirius could be traced over thousands of years by temple axes in various parts of the country.[14] The positions indicated by the axes match those for rising and setting positions of bright stars determined by modern astronomy through retro-calculation.

From these facts it becomes clear that the ancient Egyptians were fascinated—one might almost say hypnotized—by the stars, and by the circumpolar stars in particular. Whatever took place inside the pyramids had something to do with the circumpolar stars. The pyramids and the axial temples are, of course, the greatest remains in the country. The size and number of these constructions are such that the Egyptians' interest in

*A drawing of the famous circular zodiac, dated at around 300 BC and originally in the ceiling of the Temple of Denderah. The zodiac is centred on the celestial North Pole as shown by the hippopotamus at its centre, a symbol of the polar constellations.*

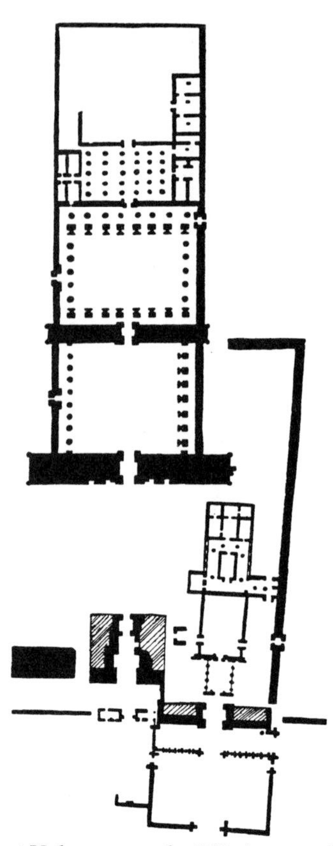

*The ground plan of temples at Medinet Habu, across the Nile from modern Luxor. The axis of the later temple, alongside and in front of the earlier one, varies by a few degrees from the original orientation. This change enabled the builders to observe the rising position of the same star sighted through the earlier structure, after the star's rising position had shifted on the horizon.*

the heavens begins to assume the proportions of some overriding compulsion, as if the solution to the deepest mystery of their civilization lay hidden in the stars.

The more we study the remains of the ancient Egyptians, the more they seem a star-struck people. As numerous as the star alignments in pyramids and temples are, they left additional representations which confirm the hints derived from astronomic orientations. In their temples we find remarkable zodiacs, the style and symbolism of which may at first seem strange and fanciful to us today, but which organize and record information that was the product of long and careful observation. The most famous such zodiac is the round zodiac of Denderah. Although produced at a late date (around 300 BC) this zodiac again reflects the pattern of the cosmic churn on Senusert's monument, the axial associations of the Djed column, and the alignments of the pyramid passages; it is centred on the north celestial pole. Near the centre of the zodiac we find the Thigh, a Hippopotamus and a Jackal, which correspond to our present constellations of the Great Bear, the Dragon and the Little Bear.[15] From the latitude of Egypt these constellations were then circumpolar.

And so from the oldest to the latest remains of ancient Egypt, a constant theme emerges — an almost compulsive interest in the stars and in the

*Like the designers of the Denderah Zodiac 2000 years earlier, Medieval Arabic astronomers continued to produce star maps centred on the celestial North Pole which is here surrounded by the constellation of the Dragon.*

circumpolar stars in particular. The story of this fascination is not restricted to the great official monuments of the country; it leads into the tombs and even inside the very coffins of the ancient Egyptians. And it leads as well to the oldest text on earth.

CHAPTER FOUR

# STAR MAPS

## The Egyptian Cosmology of Life After Death

As much as any people of any time the Egyptians regarded life on earth as a mere stepping stone, a temporary abode, in an existence which was to continue for millions upon millions of years. It is common knowledge that a considerable portion of their literary and artistic remains depict preparations for and travels in the next state of existence. It is intriguing that a people as accomplished as they were in dealing with the material world and, as has often been remarked, who had thoroughly practical and concrete imaginations, should have had such utter confidence in a life in another world. Indeed, their confidence was so great that they produced tables, charts and maps of the heavens to guide them on their way. As we follow the trail of their star maps, not only are we led to the secret of their confidence in the reality of these other worlds, but find truly astonishing evidence that their journeys into these realms were not restricted to the dead!

The volume of Egyptian remains concerning the afterlife is great. Not all Egyptians aimed for the same destination in the beyond, and their conceptions of the cosmos did not remain constant over the thousands of years of the life of their civilization. As with Senusert's monument, we are often confronted with many strata of meaning. However, in sorting out the symbols and patterns the Egyptians have left us, we gain valuable clues to the geography of their cosmos, a geography which seems to be the prototype of a universal pattern.

Many authors refer to the Egyptian world beyond this one as the 'Underworld', which may lead one to imagine that it is beneath the surface of the earth. The Underworld was not underground. This is made clear in a late and rather degenerate illustrated composition known as the *Book of Pylons*. In this work the Underworld or 'Duat' is identified as the course of the sun through the 12 hours of the night. Each hour was conceived as separated from the next by a pylon — hence the name of the work.

The Egyptologist Wallis Budge tells us that the Duat or Underworld was a great valley with mountains along each side and a river at the bottom. The mountains on one side separated the Duat from the earth, and those on the other separated it from heaven.[1] It was, then, in the sky between the earth and the outer heavens. This is made clearer in the many cosmograms showing the sky goddess Nut uplifted from the earth god Geb by the god of the air, Shu. Nut's body is typically spangled with stars. She was said to swallow the sun at day's end and to give birth to it again each morning. The Valley of the Duat was in the body of Nut.

Although this valley was in the domain of Osiris and Ra, it contained monstrous serpents and demons hostile to the spirits of men. It has been suggested by many Egyptologists that it is the same valley as that later described by the psalmist: 'Yea, though I walk through the valley of the shadow of death, I will fear no evil: for thou art with me; thy rod and thy staff they comfort me.' (PSALM 23)

Despite its unsavoury inhabitants, the souls of the blessed dead who followed the instructions of the *Book of Pylons* hoped to become companions of the sun and to journey in the boat or barque of Ra as he sailed on the river of the Duat. Under the protection of Osiris and Ra, and with the aid of the formulas and prayers the priests had given them, they could pass the guardian monsters along their path.

If it seems strange that the Egyptians thought of the sun travelling in a boat, it may be helpful to remember that all the functions served by modern highways and airports today were in Egypt served by the Nile. It was natural for the Egyptians to associate travel of almost every kind with boats. It was not just the sun god who travelled in a boat; the moon, planets, certain constellations and stars are depicted in boats as well. These boats automatically conveyed the idea that the symbolized beings and spheres were in motion. Curiously, the Peruvians depicted the moon travelling in a boat. It is curious, too, that at Gizeh, Saqqara and other sites there are around some pyramids ancient boat pits excavated out of the bedrock and constructed from brick. Some of these are aligned east-west and others north-south. Commonly known as 'solar boats', the name may

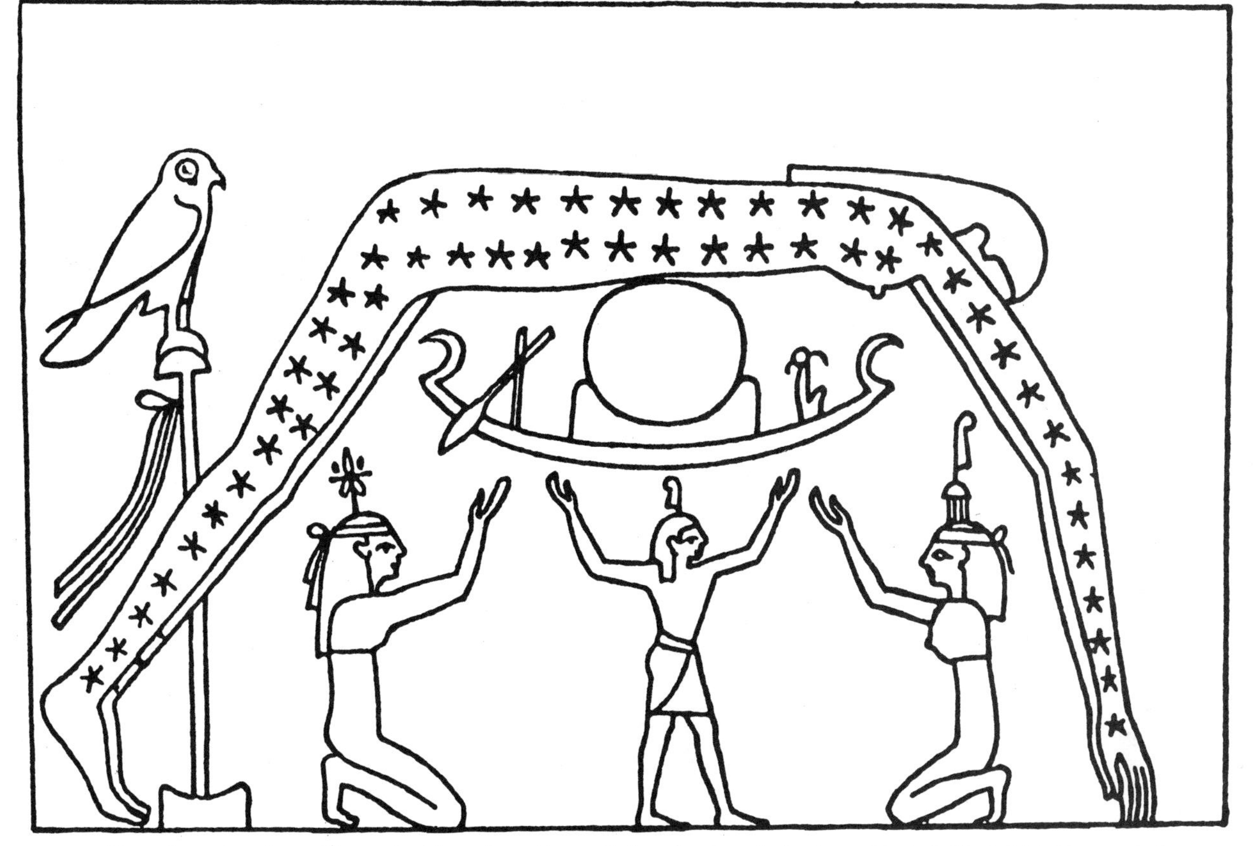

*Nut, Egyptian sky goddess, stretches above the earth from horizon to horizon. She is held aloft by Shu, god of the air, seen here in the centre with upraised arms. Above him is a boat carrying the sun. Attendant goddesses flank Shu. The drawing is from a late New Kingdom papyrus about 1100 BC.*

*The god Tem or Atum seated in his boat; the boat rests upon a symbol for the vault of heaven, indicating that this is a celestial boat.*

*The sky goddess Nut, the vault of heaven, with a star-spangled body and the discs of the sun and moon.*

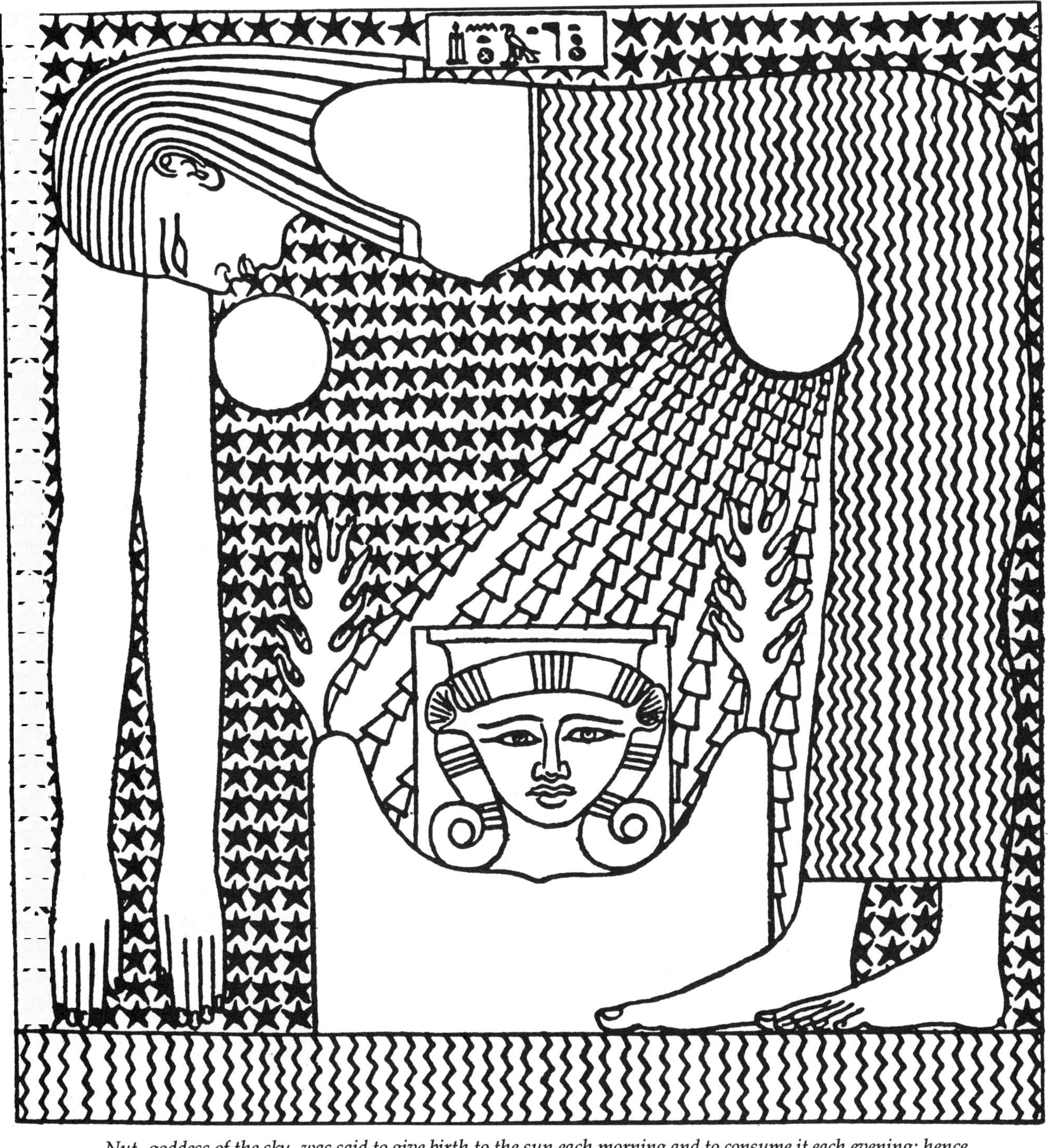

*Nut, goddess of the sky, was said to give birth to the sun each morning and to consume it each evening; hence one orb was placed near her womb, a second near her mouth. Here the sun's rays bring forth the goddess Hathor, whose head rests on the hieroglyph symbolizing the horizon. From a New Kingdom decorative panel.*

*A hawk-headed god, probably Ra, holds the symbol of life sailing in a boat with the sun and moon. The boat rests upon a symbol for the vault of heaven.*

*The falcon-headed god Ra holds an* ankh, *the symbol of life, while sitting in a boat of the sun, surmounted by the solar disc. The boat sails toward a star field.*

*Remen-heru-an-Sah, ruler of the 32nd decan, a ten degree section of the sky.*

*The boat of Ra, here symbolized by a disc upon the head of a falcon. Over the bow and stern are* utchats *or symbolic eyes. The boat sails upon the stars of heaven. From the Papyrus of Nu.*

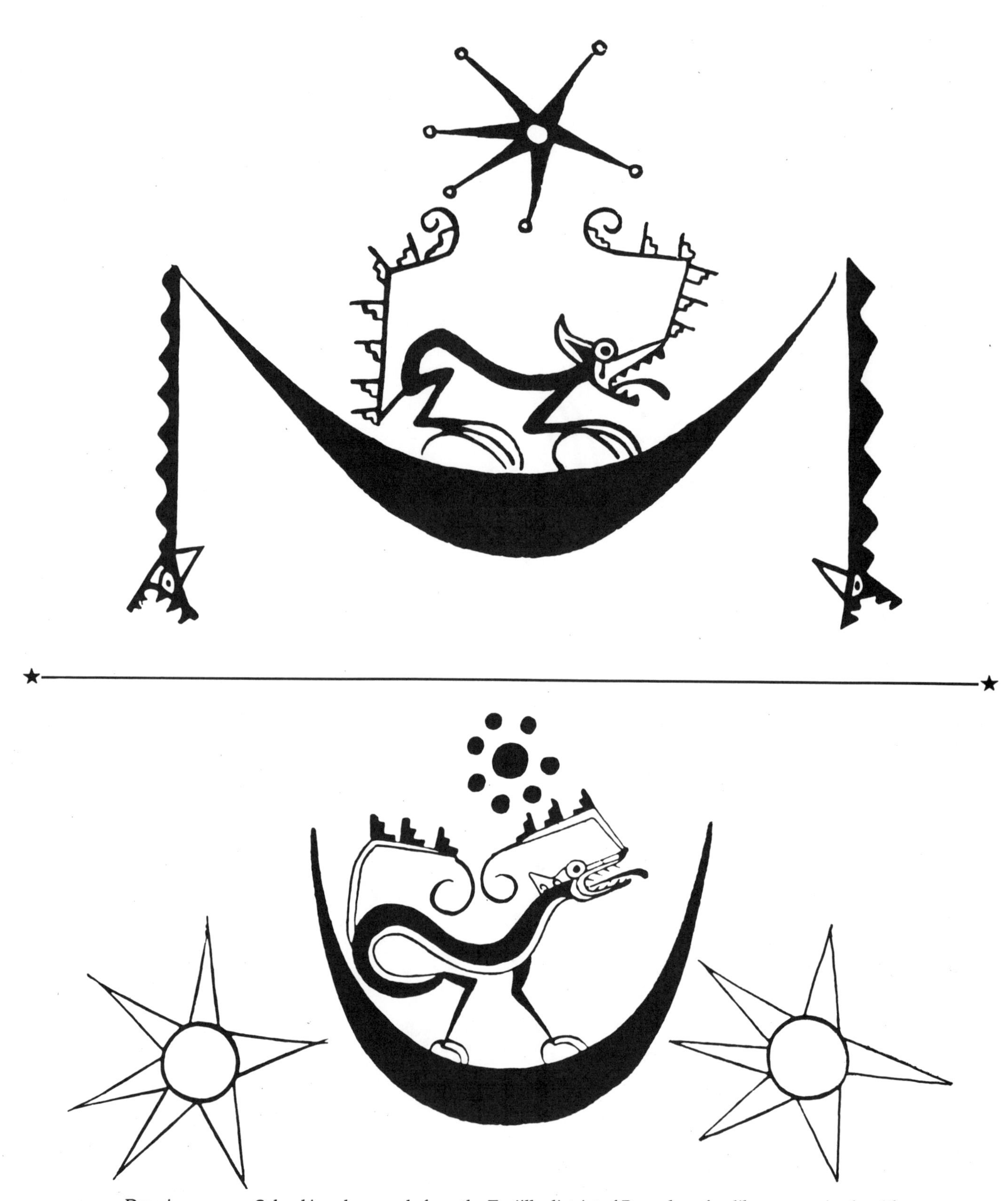

*Drawings on pre-Columbian clay vessels from the Trujillo district of Peru show fox-like moon animals with designs attached to their heads and tails resembling the stepped throne which often crowns the Egyptian goddess Isis.*

*One of the boat pits alongside the Great Pyramid.*

*The Egyptians were not the only people who pictured the celestial bodies as gods and men travelling the heavens in boats, as is evident in this image from Mesopotamia.*

be appropriate to those that parallel the path of the sun but not to those pointing north; these might, by inference, be called 'stellar boats'. About them little is known, aside from the surmise that they were the scenes of rites in some way connected with the heavenly bodies.

At any rate it is clear that in the *Book of Pylons* and in a companion piece, the *Book of that which is in the Underworld*, written around the same time, we are dealing with a geocentric conception of the universe: the course of the sun goes around the earth. However, both these compositions were late attempts (around 1600 BC) by the Theban priesthood to accommodate the much more ancient *Book of the Dead* to the pattern of a strictly solar cult. They were neither conceptually nor popularly successful. As we find so often in the study of ancient peoples, their latest conceptions were a debasement of earlier beliefs.

What we now know as the *Book of the Dead* was called the *Chapters of Coming Forth by Day* by the Egyptians. Its modern title is due to the circumstance that copies of it were interred with the deceased. These copies date back no further than the *Book of Pylons*, but the scribes who recorded these latest versions of the *Book of the Dead* declared that parts of it went back to times that were ancient even for them.[2]

In the *Book of the Dead* a much more sophisticated cosmology emerges. There is of course the familiar judgment scene before the gods Osiris, Thoth and Anubis which was the invariable fate of everyone. Those who passed this judgment could go to places called Sekhet Aaru, the Field of Reeds, and Sekhet Hetepet, the Fields of Peace or the Elysian Fields, and other similar realms. In the papyri these Elysian Fields are shown as a group of islands separated by streams or rivers with the whole group surrounded by water. A common feature of the Elysian Fields is that they

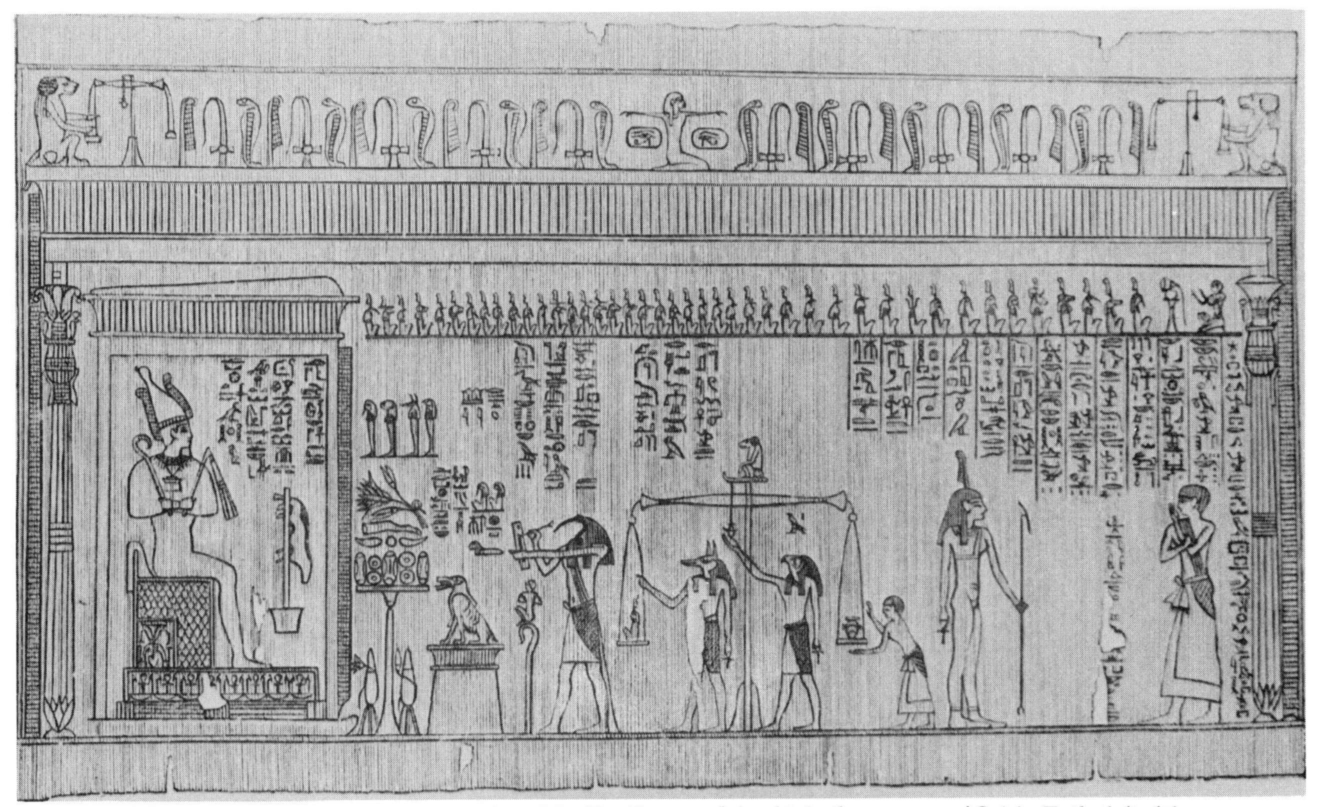

*Egyptian judgment scene: The heart is weighed by Horus and Anubis in the presence of Osiris. To the left of the scales Thoth records the result. To the right, the goddess of truth holds the symbols of life and death. From a New Kingdom papyrus (1500-1000 BC).*

contain a boat in which there is a flight of steps. This boat is found at the end of a canal, and the steps likely indicate that this was the point of entry after one had 'risen up' to gain access to this celestial realm.

These maps of paradise identify the person for whom they were executed and show him engaged in occupations such as offering incense to the gods, paddling a boat, smelling a flower and ploughing with oxen. At first this may seem a rather small scale paradise, but the symbolic nature of the images becomes apparent when we are told, as in the Papyrus of Nebseni, that the stream by which Nebseni is ploughing is 1000 measures in length and the width of it 'cannot be said'. Furthermore, this stream is unlike any on earth for 'in it are neither fish nor worms'.[3]

As we shall see, the mythologies of many peoples refer to the Rivers of Heaven, and these rivers are usually associated with the circuits of the planets. This identification emerges partially from the fact that these rivers are often seven, eight or nine in number. It is significant, then, that in the Papyrus of Nu the divisions of the Underworld, which in other Egyptian texts are called fields, cities or halls, are there described as seven mansions or circles each with a doorkeeper, a watcher and a herald.[4] We can provisionally conclude that just as followers of the *Book of Pylons* hoped to accompany the sun, the older and larger cosmology included the planets as the abodes of the dead.

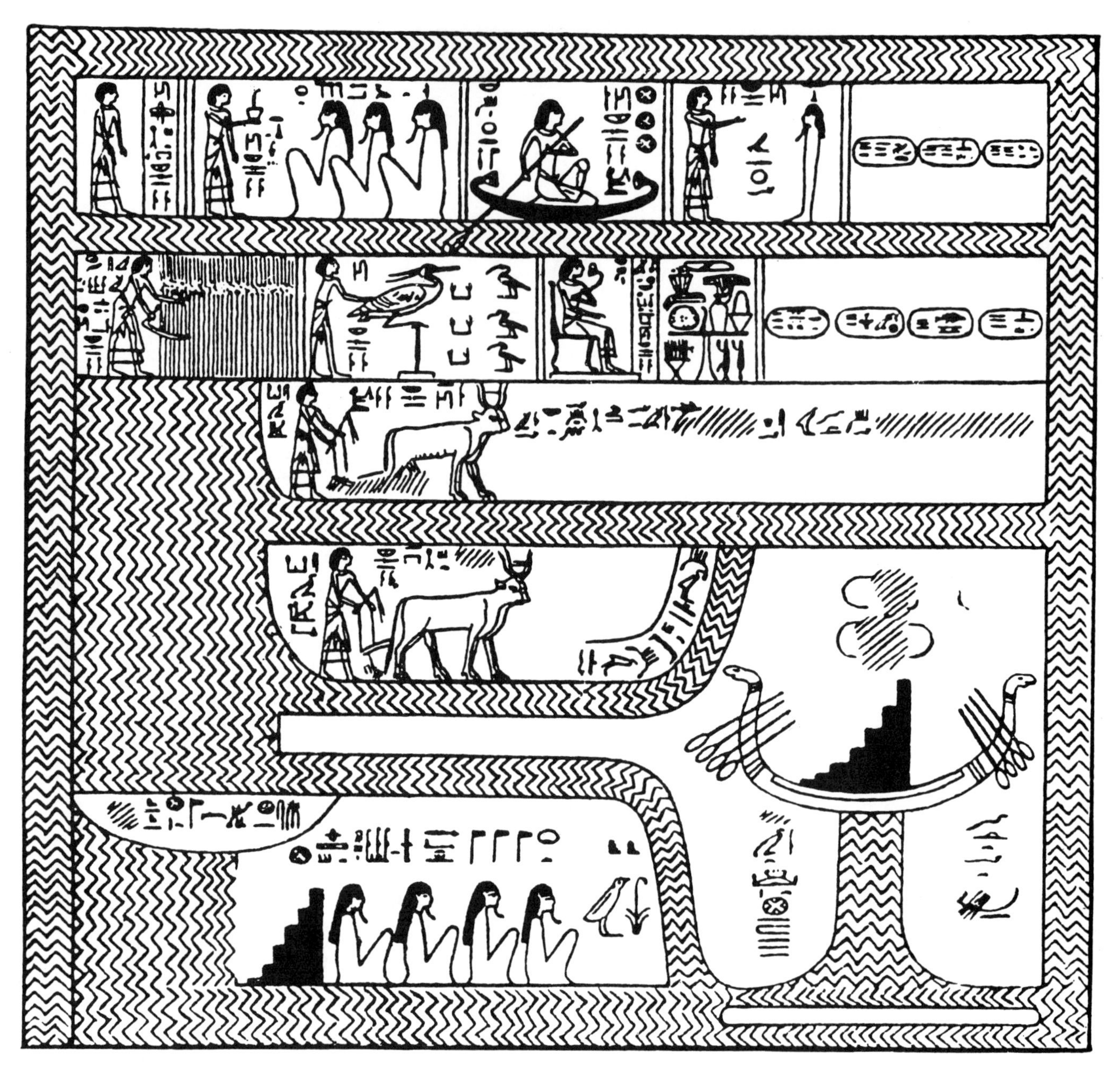

*The Egyptian Elysian Fields (called Sekhet Hetepet) as shown in the Papyrus of Nebseni, a late papyrus of the* Book of the Dead. *The Egyptians depicted water with zig-zag lines, indicating that they thought of the other world as islands separated by rivers and canals. The papyrus is organized in panels. Starting at the upper right, Nebseni enters the Elysian Fields and is identified as the scribe and artist of the Temple of Ptah. Second panel: Nebseni makes an offering of incense to the 'great company of the gods'. Third: Nebseni sits in a boat paddling; above the boat are three symbols for 'city'. Fourth: Nebseni addresses a bearded mummy figure. Fifth: Three pools or lakes called Urti (or Hemat), Hetep (or Hast) and Qetqet. Sixth: Nebseni reaping heavenly food. Seventh: Nebseni grasps the Bennu bird which is perched upon a stand; in front are three* kas *and three* khus. *Eighth: Nebseni sits and smells a flower; the text reads: 'Thousands of all good and pure things to the* ka *of Nebseni.' Ninth: A table of offerings. Tenth: Four pools or lakes called Neb-taui, Uakha, Kha (?) and Hetep. 11th: Nebseni ploughs with oxen by the side of a stream which is one thousand (measures) in length, and the width of which 'cannot be said; in it there are neither fish nor worms'. 12th: Nebseni ploughs with oxen on an island 'the length of which is the length of heaven'. 13th: A division shaped like a bowl in which is inscribed: 'The birthplace of the god of the city, Qenqen (et nebt).' 14th: An island whereon are four gods and a flight of steps. The legend reads: 'The great company of the gods who are in the Elysian Fields.' 15th: The boat Tchetetfet with two sets of four oars, possibly the four rudders of heaven, floating at the end of a canal. In it is a flight of steps, symbol of ascent. The place where it lies is called the Domain of Neth. Above the stairs in the boat are two pools, the names of which are illegible.*

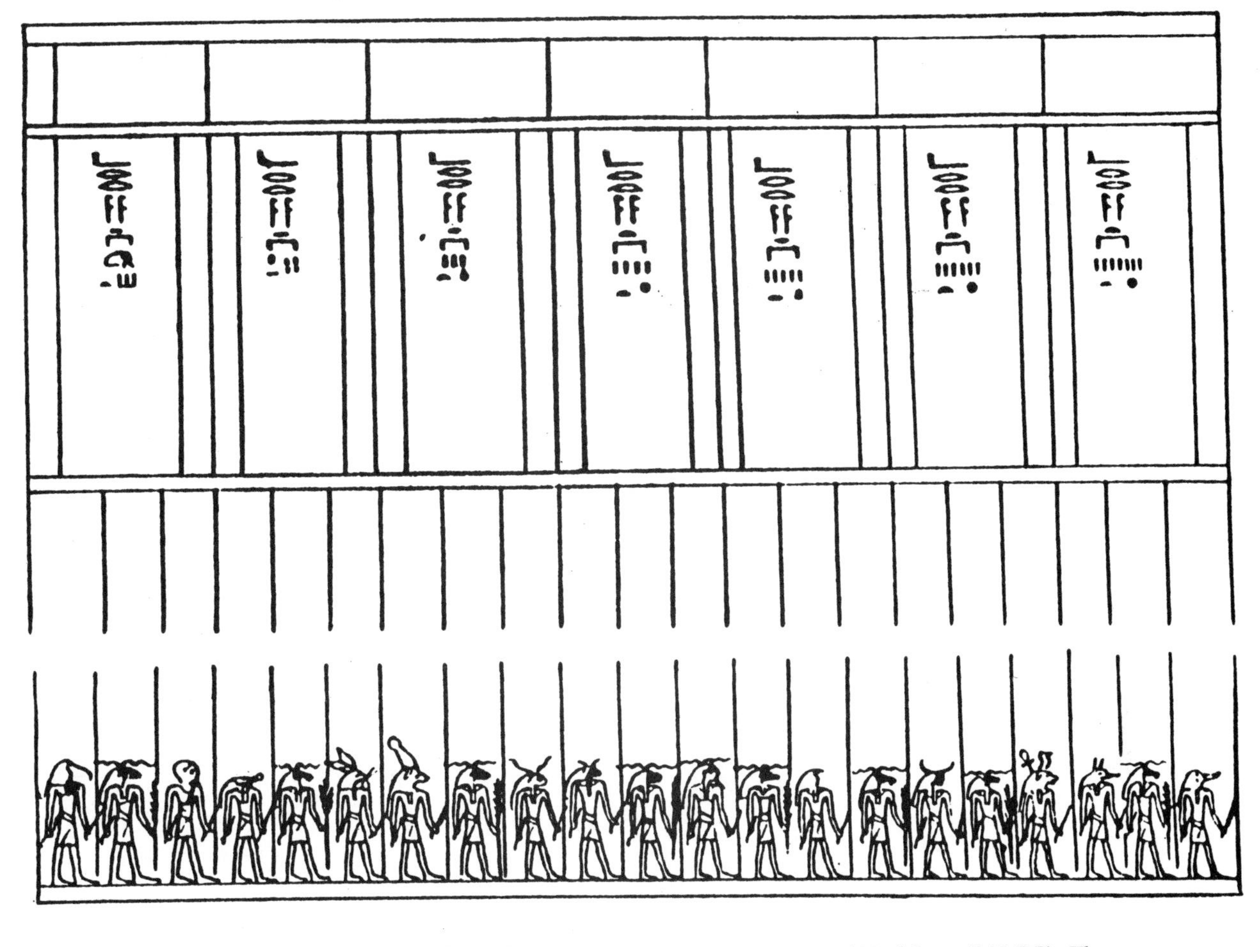

*The seven mansions or circles of the Underworld as shown in the Papyrus of Nu (about 1000 BC). The seven circles, each with a doorkeeper, a watcher and a herald, correspond to the circuits of the planets.*

The ancients referred to both stars and planets as 'stars', but there is no doubt that the Egyptians were well aware of the distinguishing motions of the bodies in the solar system. A number of sources make this clear, among them the most elaborate star maps of the Egyptians. A particularly fine example of such a map is found in the tomb of Senmut (about 1473 BC). (Senmut was Queen Hatshepsut's architect who built her strikingly individual temple at Der el Bahri. His ability as a designer seems to have carried over into his tomb.) The tomb of pharaoh Seti I (who reigned 1326–1300 BC) in the Valley of the Kings contains another. Appropriately, these large star maps cover the ceilings of these tombs. In these star maps and in other astronomical reliefs on the ceilings of various temples, such as the Ramesseum erected by pharaoh Ramses II (1290–1223 BC) at Luxor, the planets are identified and described.

Venus was called 'the star which crosses' or 'the crosser', a reference to its appearing both in the morning and evening. Mars was sometimes called

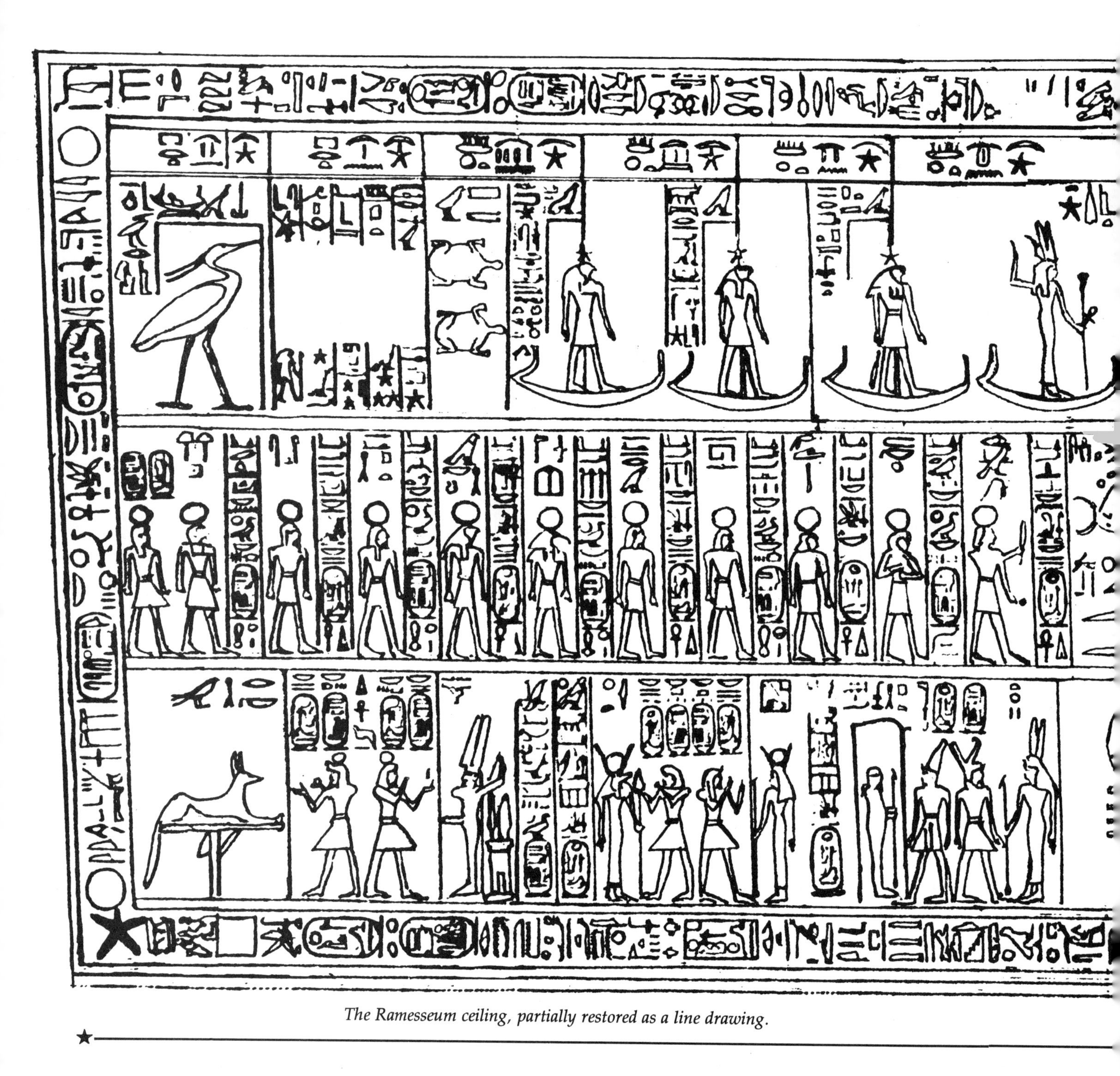

*The Ramesseum ceiling, partially restored as a line drawing.*

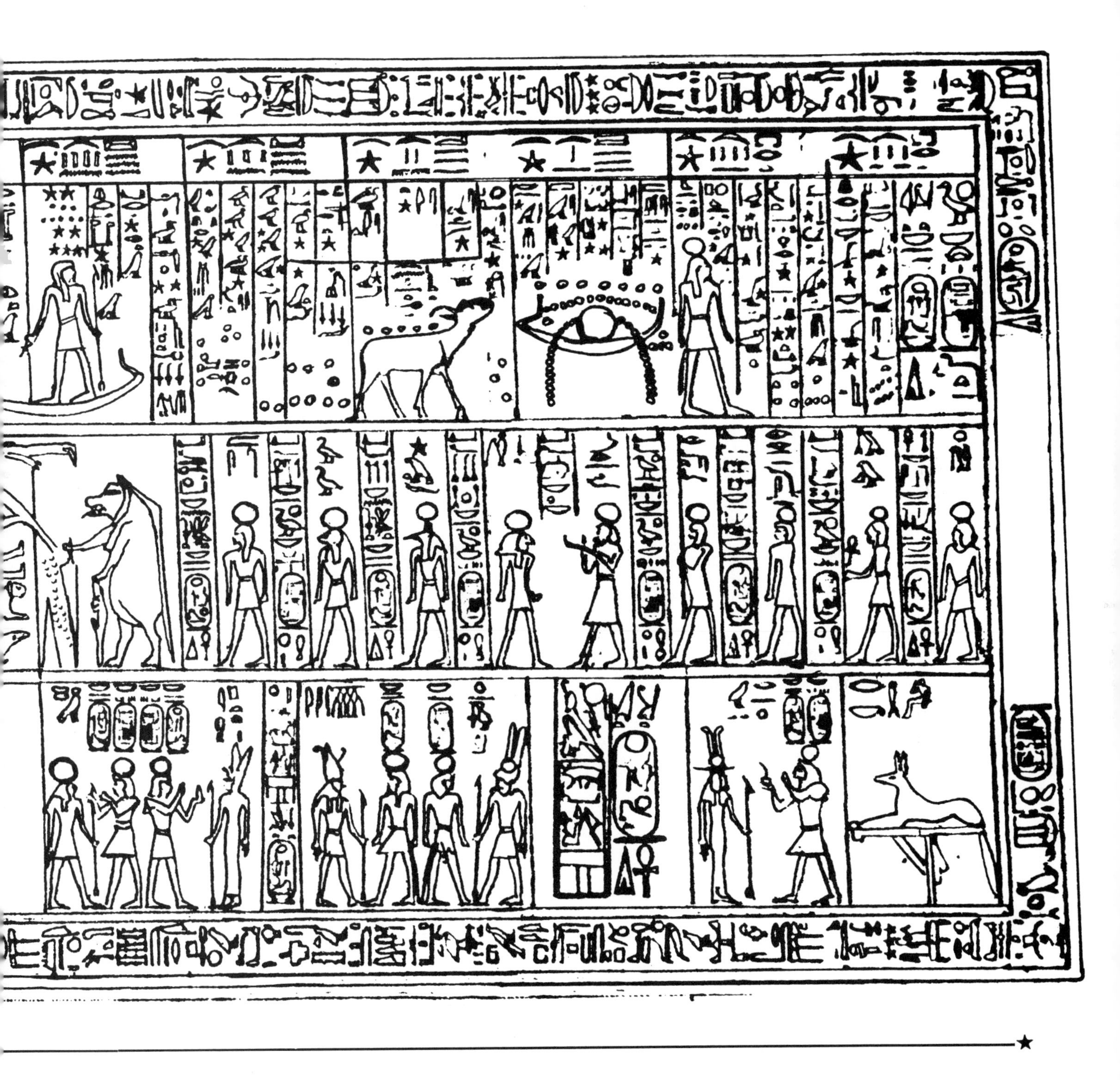

*The practice of inscribing astronomic information inside coffin lids persisted for over 2000 years and well into the Christian era as shown in the coffin lid of Heter, AD 125. The central figure is a variation of the sky goddess Nut. Between her head and arm on the right are Orion and Sirius. The West Wind is symbolized in the upper left corner, the North Wind is upper right, the East Wind in the lower left corner, and South Wind in the lower right. The lower register on the left has the hours of the day; the corresponding register on the right has the hours of the night. In the upper left register are the northern Constellations, and several deities symbolized by the Baboon (Thoth), the Falcon (Horus) and four sons of Horus. In the upper right register are six figures, five of which are plantes; however the name columns were left blank above them and other figures, preventing precise identification. Of unusual interest are the Greek signs of the zodiac surrounding the body of Nut, above, and among which, on the left side are recorded the planetary positions at the birth of the coffin's owner, thus forming a horoscope dating to the early part of October AD 93. A demotic inscription elsewhere on the coffin gives the owner's age at death as 31 years, 5 months and 25 days, allowing an unusually exact dating of the coffin.*

*A portion of the star chart found inside a coffin of the ninth or tenth dynasty (before 2000 BC). The chart records the shifting rising times of stars as they vary throughout the year. The figure of a thigh on the right represents the Bull in the northern constellations.*

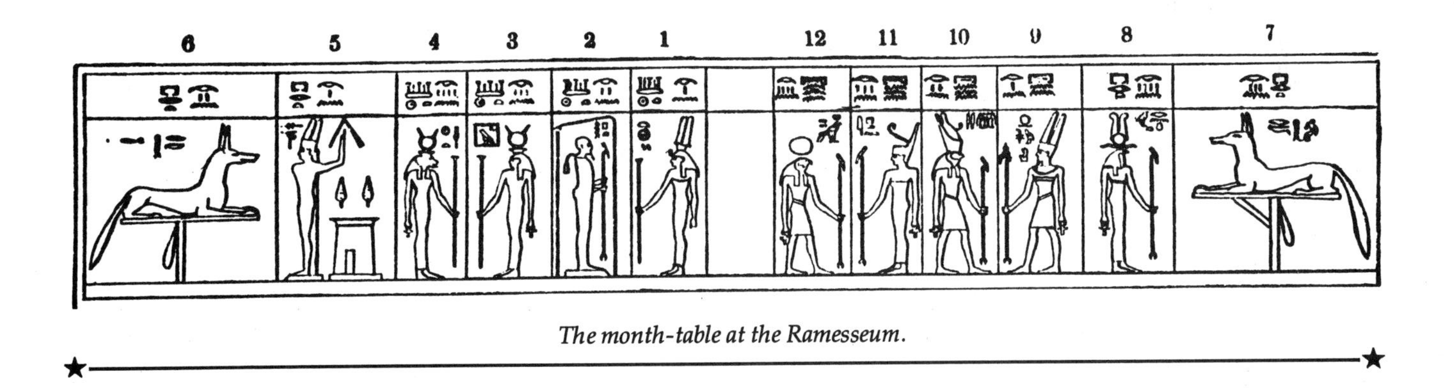


*The month-table at the Ramesseum.*

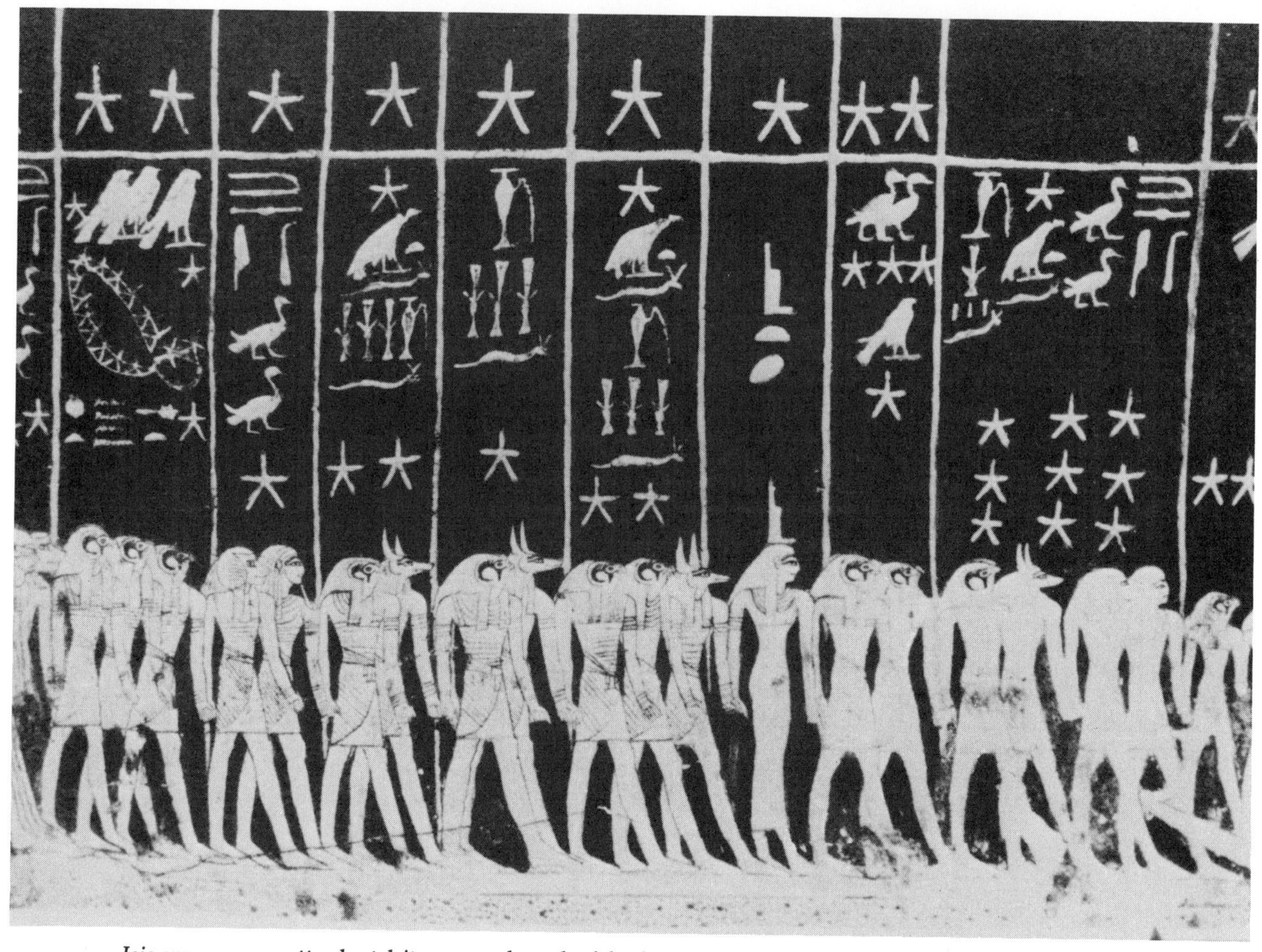

*Isis appears as an attendant deity among the gods of the decans on the ceiling of the tomb of Seti I. Isis, centre, is shown crowned by a small step-like throne and another throne above her. Similar symbols appear in early Peruvian astronomic art.*

'the star which journeys backwards in travelling', a description of the retrograde motion that is particularly noticeable with this planet. All the planets from Mercury to Saturn were known, and Mercury too was recognized as a body appearing both in the morning and the evening. Mars, Jupiter and Saturn were all seen as aspects of Horus; Mars was 'the red Horus', Jupiter 'the Horus who illuminates the two lands', and Saturn was 'Horus, the bull of the sky'.[5]

The large star maps in tombs and temples record an amazing variety of information other than planetary designations. They are composite representations of a variety of astronomical lore and techniques, organized partly by aesthetic criteria into a single visual field. Much of the information in these visually elaborate sources is concerned with the rising times and culminations of stars.

As we saw with the stellar alignments in temples noted by Lockyer, the Egyptians had been watching the risings of stars for centuries before the development of these large star maps. There are, in addition, 12 known cases of decanal star clocks inscribed *inside* coffin lids, dating from about 2160 to 1786 BC. A decanal star clock is a diagramme naming stars or groups of stars as they rose throughout the year.[6] Knowing the date, an observer of the sky could use the star clock to calculate the hour of the night. Conversely, knowing the hour and date, he would know which stars were due to rise.[7]

Decanal star lists occupy good portions of the maps of Senmut, Seti I and the astrological ceilings in temples. These maps also depict major, easily identifiable stars such as Sirius and the constellation of Orion, lunar calendars, certain deities, (and sometimes even images of the king, if the work happened to be commissioned by Ramses II, who was forever making clear how important he was). In later temples and cosmograms, the signs of the zodiac also appear.

With all of this information, however, it is significant that the large star maps depict the circumpolar constellations again and again in positions of central importance. Even in the rectangular fields of these compositions, everything seems to revolve around the Hippopotamus, Thigh or Bull, and Crocodile which represent the northern constellations. It is interesting that on the ceiling of the Ramesseum they are positioned directly above a Djed column, which is itself in the centre of the northern register of the ceiling—still another illustration of the Djed as the axis or pole.

No one pretends to understand these star maps completely.[8] We do know that the motions of the sun, planets and ecliptical constellations were certainly important to the Egyptians. They carefully studied the risings of the stars and the circuits of the moon. We can see that even in the context of these other concerns the dominant importance of the circumpolar stars—the stars to which the pyramid passages point—runs straight through their entire civilization. Curiously, in sorting out all the astronomic associations of these images, scholars tend to lose sight of the fact that many of these maps and charts are found in tombs and inside coffin lids. Given the Egyptians' practicality, the appearance of these maps in tombs directly suggests that they expected to make use of them in reaching higher realms—just as those who carried the *Book of the Dead* into the grave expected to make use of it. Indeed, the very texts written on the walls of chambers inside some pyramids describe journeys to the circumpolar stars in particular, and in the most detailed account which survives we can follow step by step the travels of an individual named Unas, a pharaoh of Egypt's Old Kingdom. However, in the itinerary of his celestial wanderings, the condition of Unas is documented with precise and thorough detail; the texts repeatedly make the extraordinary assertion that 'this Unas has not died'.

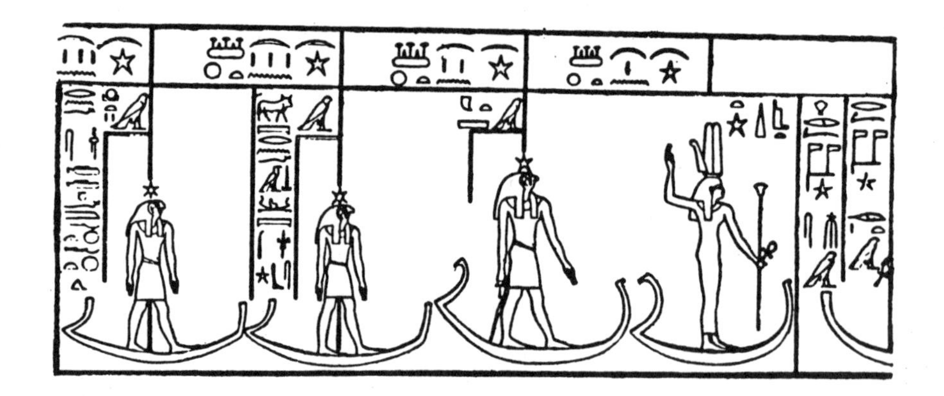

*Planets as shown on the Ramesseum ceiling: from the left, Mercury, Saturn and Jupiter stand in their barques following Sirius, represented here by Isis, crowned and also in a barque. (In this 1850 drawing, the artist has given the stars over Mercury and Saturn six points when they should have only five.)*

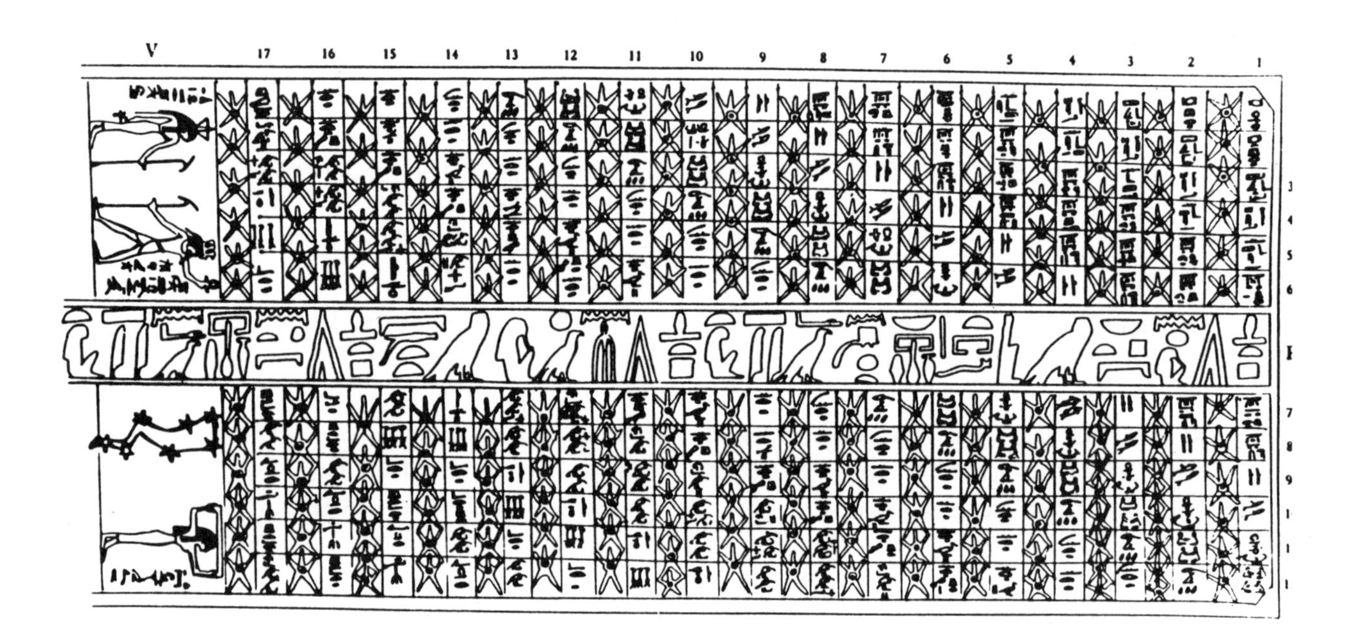

*Right half of the inner face of an 11th dynasty coffin lid, 2061-2011 BC; with the inscription 'King's wife, King's sole ornament, Priestess of Hathor'. The coffin lid records the rising positions of stars at ten day intervals throughout the year. The practice of inscribing astronomic information inside coffin lids persisted well into the Christian era. One of the latest examples known dates to AD 125.*

*A portion of the elaborate ceiling of the Outer Hypostyle Hall of the Temple of Hathor at Denderah dating from AD 14-37. This ceiling records an amazing amalgam of astronomic concepts. The female figure with a star above her head at upper left represents the first hour of the night; next is the symbol for Capricorn; the Falcon above Capricorn is identified as Jupiter; the next goddess with a star is the second hour of the night; then the Hippo, Bull and Falcon-headed Spear-thrower depict the Circumpolar Constellations followed by the third hour of the night and Sagittarius. The lower register has gods in boats symbolizing decans.*

*The northern polar constellations as shown on the tomb ceiling of Pharaoh Seti I at Thebes. The Hippopotamus with a Crocodile on its back probably represents Draco the Dragon. The full figure of a bull is elsewhere represented simply as a head or thigh of a bull.*

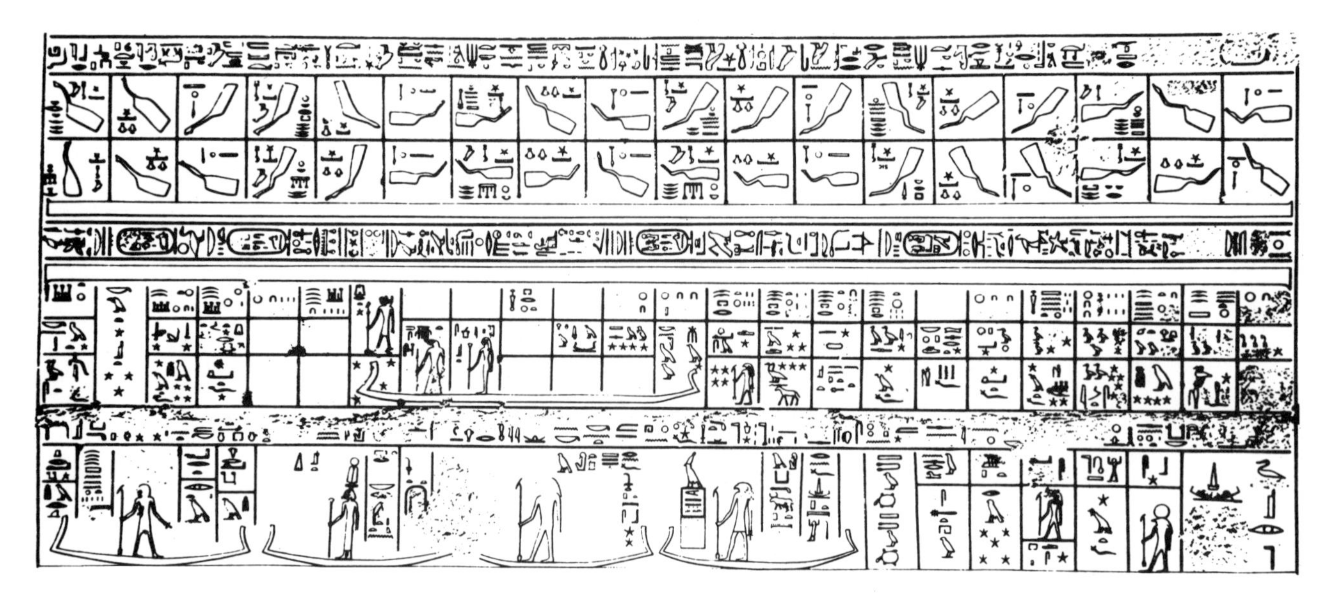

*The inner face of a lid from a granite sarcophagus from Kom Abu Yasin, dated to the 30th dynasty (about 359–341 BC). The lower registers record the hourly decans, planets and attendant deities. It is especially noteworthy for the upper register which records monthly variations of the Polar constellation usually represented by the Bull. Here only the foreleg of the Bull symbolizes the constellation as it rotates around the pole. The accompanying text equates Osiris, god of the dead and of resurrection, with this heavenly Bull.*

*Detail from the Ramesseum ceiling showing the Djed column in the centre of the northern register. Above the Djed are the northern constellations, identified by the Hippopotamus, the Crocodile and the Bull.*

CHAPTER FIVE

# THE STAR WALKER

## The Oldest Text Known on Earth

Those things preserved in Egypt, Plato wrote, were said to be the oldest because the location of the country had largely spared it from the catastrophes and upheavals that had obliterated less favoured civilizations. Indeed, beneath the sands of Saqqara, on the walls of chambers under the small ruined Pyramid of Unas, is the oldest text known on earth. It reaches back at least 43 centuries and possibly many more. The ceilings of these chambers are covered with stars and in the text itself, as well as in others in nearby pyramids slightly more recent, are numerous references to individuals who travelled to the stars — and *returned*.

Like most of the pyramids at Saqqara, the Pyramid of Unas now looks like a pile of gravel. Unlike the large Gizeh pyramids, constructed throughout with massive well-cut blocks of stone, most of those at Saqqara were built with relatively small blocks arranged as a series of walls. The spaces between the walls were generally filled in with rubble, so that when through the ravages of time or the hand of man, the original casing disappeared, the pyramid rapidly lost its geometric shape. By 1500 BC some of its stones had been used in another monument. Although the pyramid he left us is now anything but grandiose, the texts Unas had recorded in the chambers beneath it provide more food for thought than many of the more spectacular constructions before and since.

*The ruined mound of the Pyramid of Unas at Saqqara.*

The pyramid was opened and the texts discovered by Gaston Maspero in 1881. The rather small 'burial chamber' was found empty except for a large, stone, lidded sarcophagus — which also was empty. Of greatest interest, however, are the texts on the walls of this and connecting chambers. Generally speaking, Egyptologists have a firm command of the hieroglyphic symbols with which the Egyptians recorded their language during the phase of their history which is relatively certain; that is from about 1567 BC on. But the symbols employed and the rules of the language did not remain constant. In earlier times the language was in such an archaic form, and remnants of it are now so rare, that much of their meaning — sometimes all of it — defies translation.

In the case of the Text of Unas, Egyptologists have been able to supply words or at least syllables for almost all of the hieroglyphs. The text is organized into sections called 'utterances' and 'spells'. Gaston Maspero first attempted a translation in 1882 but despite nearly a century of such efforts, there is still no general agreement among scholars as to where the text begins or ends. Nor is it even clear whether it is all one text or a series of texts. In addition, as they have been translated and presented, the texts sometimes yield statements and references so disjointed, bizarre and remote from our own world that their meaning remains elusive. Even in the context of surrounding material, some utterances are no less opaque than when presented in isolation:

The one bitten by Atum has filled the mouth of Unas
while he twisted (himself) a twisting.
The centipede is struck by the inhabitant of the house.
The inhabitant of the house is struck by the centipede.
That lion (should be) inside this lion.
The two bulls fight inside the Ibis.[1]

Sometimes we seem to have descriptions of a phantasmagoria:

A face falls on a face, a face has seen a face.
The mottled knife, black and green,
went forth against it:
it has swallowed that which it tasted.[2]

And sometimes, one would think, either the original scribe or the translator had an outrageous sense of humour: 'Thou art to rape the two holes of the stone door jamb!'[3] Undaunted, a German scholar has suggested in all seriousness that this refers to a sexual act to fertilize the pyramid![4]

In spite of such places where the translation seems to mingle sense with nonsense, large sections of the text are more coherent and a probable order for some of them does emerge, suggested in part by the pattern of the chambers themselves. The texts occur in the Entrance Passage, the Antechamber, the passage to the Sarcophagus Chamber and in the Sarcophagus Chamber. It is possible that some texts in the Entrance and Antechamber were meant to be read as the priests *left* the pyramid after the performance of rituals in the Sarcophagus Chamber, but others seem to be clearly introductory. In the Entrance Passage itself we find that Unas is described almost as a god:

Unas causes the grass to become green on the two banks of the horizon.
Unas brings the green brilliance to the Great Eye which resides among the green pastures.
Unas takes his seat which is on the horizon.
Unas arises as Sebek, the son of Neith.
Unas eats with his mouth, Unas urinates, Unas cohabits with his phallus,
Unas is the lord of seed, he who takes the women from their husbands,
wherever Unas wants, according to the wish of his heart.[5]

Naturally, the most important texts occur in the Sarcophagus Chamber and, of course, the standard interpretation of these by Egyptologists is that they are rituals and formulas for the guidance of the soul of the dead king. But that is not what the texts themselves say. In the Sarcophagus Chamber itself, in the longest utterance of all in this oldest text on earth, it is repeated no less than *twenty-four* times that 'He is not dead, this Unas is not dead.

*The Text of Unas on the south wall of the sarcophagus chamber.*

The utterance begins:

Atum, this thy son is here,
Osiris, whom thou hast preserved alive — he lives!
He lives — this Unas lives!
He is not dead, this Unas is not dead:
he is not gone down, this Unas is not gone down:
he has not been judged, this Unas has not been judged.
He judges — this Unas judges![6]

Atum was one of the names for the High God or the Great God of the Egyptians. In the second line Unas is identified with Osiris, the god and judge of the Egyptian dead, and, more precisely, the god of resurrection. The next part of the utterance continues: 'Shu, this thy son is here' and then the rest of the formula is repeated as above. Shu was god of the air, and with the other chief intelligences or divinities of the Egyptians such as Tefnut, Geb, Nut, Isis, Set, Nephthys and Thoth substituted one by one in place of Atum and Shu, the same formula is repeated two dozen times.

Taking for granted that this is only a funeral text, scholars do not dispute the repeated statements that Unas has not died. It is, they tell us, his soul that still lives. However, it is far more difficult to explain how it is that 'he has not been judged, this Unas has not been judged', since after death judgment before Osiris and the other gods was the fate that awaited *everyone*, including the king.

This longest of the utterances in the texts of Unas begins on the south wall of the Sarcophagus Chamber and continues on the east wall. It is via the passage through the east wall that one must enter or leave the chamber. It is interesting that on this east wall near the passage this long, one might say hypnotic, utterance ends:

He is not dead, this Unas is not dead:
he is not gone down, this Unas is not gone down:
he has not been judged, this Unas has not been judged.
He judges — this Unas judges!
Thy body is the body of this Unas,
thy flesh is the flesh of this Unas,
thy bones are the bones of this Unas.
Thou goest, this Unas goes:
this Unas goes, thou goest.[7]

In the beginning line of the next utterance on the same east wall, it seems a special passage has been opened:

The doors of the horizon open themselves, its bolts slide.[8]

*View from inside the sarcophagus chamber of the Pyramid of Unas. Texts surrounding the door describe the out-of-body departure, flight and return of Unas.*

Unas takes off and rises up:

he comes to thee, O his father. . . . [9]
Mayest thou grant that this Unas seize the Cool Region. . . . [10]
(Thou) standest (as king) over the places of the primeval ocean . . .
thou risest with the father Atum.[11]
Thou goest up and openest thy way through the bones of Shu (the air),
the embrace of thy mother Nut (the sky) enfolds thee.[12]

Next it sounds like Unas is orbiting the earth and rising and setting like the sun, moon, planets and stars:

Thou goest up, thou goest down;
thou goest down with Re, darkened with Nedy.
Thou goest up, thou goest down;
thou goest up with Re, and thou risest with the Great Raft.
Thou goest up, thou goest down:
thou goest down with Nephthys, darkened with the Evening Barge.
Thou goest up, thou goest down:
thou goest up with Isis;
thou risest with the Morning Barge.[13]

In fact, it is said that 'Unas goes around the sky like Re', the sun.[14] And Unas is likened to a star: 'The king appears as a star';[15] 'He is the star of the Lower Sky.'[16]

*The Night Boat (left) and the Day Boat (right) sail over the body of Nut. In the Pyramid Text Unas is described journeying in these boats.*

But Unas has a specific destination. It is the place in the heavens to which the pyramid passages point:

The Father of Unas, Atum, seizes the arm of Unas and he assigns Unas to . . . the Circumpolar Stars.[17]
Thou art to purify thyself with the cool water of the Circumpolar Stars.[18]

The sun-folk shall call out to you,
for the Circumpolar Stars have raised you aloft.[19]

Yet, as much as anything, perhaps the point of Unas' journey has been this:

Thou hast become being, thou art become high,
thou art become spirit!

Cool it is for thee in the embrace of thy father,
in the embrace of Atum.[20]

We are told literally dozens of times in these texts that Unas has not died:

O Unas thou art not gone dead,
thou art gone alive to sit on the throne of Osiris.[21]
Re-Atum does not give thee to Osiris,
He reckons not thy heart.[22]

*Receiving the waters of immortality at the Celestial Tree, which is both the* Axis Mundi *and the Tree of life. Detail from the* Papyrus of Khonsu-mes A.

*Detail from a star map in the tomb of Montemhet at Luxor about 650 BC. A man holding the emblems of power over life (the* ankh *) and death (the* was *sceptre) travels in a star boat against a star field. The man himself is associated with groups of stars.*

The weighing of the heart of the deceased was an essential part of the judgment of the dead, yet again and again it is emphasized that this has not happened to Unas. In attempting to account for the fact that Unas has not been given over to Osiris, some scholars have theorized that at the time the texts were recorded there was a political conflict between the priests of Osiris and the priests of Ra. They hypothesize that the texts were written by the priests of Ra and that is why Unas is not given to Osiris. But this political interpretation actually takes us nowhere because Unas himself is identified with Osiris; in terms of the ritual, he impersonates Osiris.

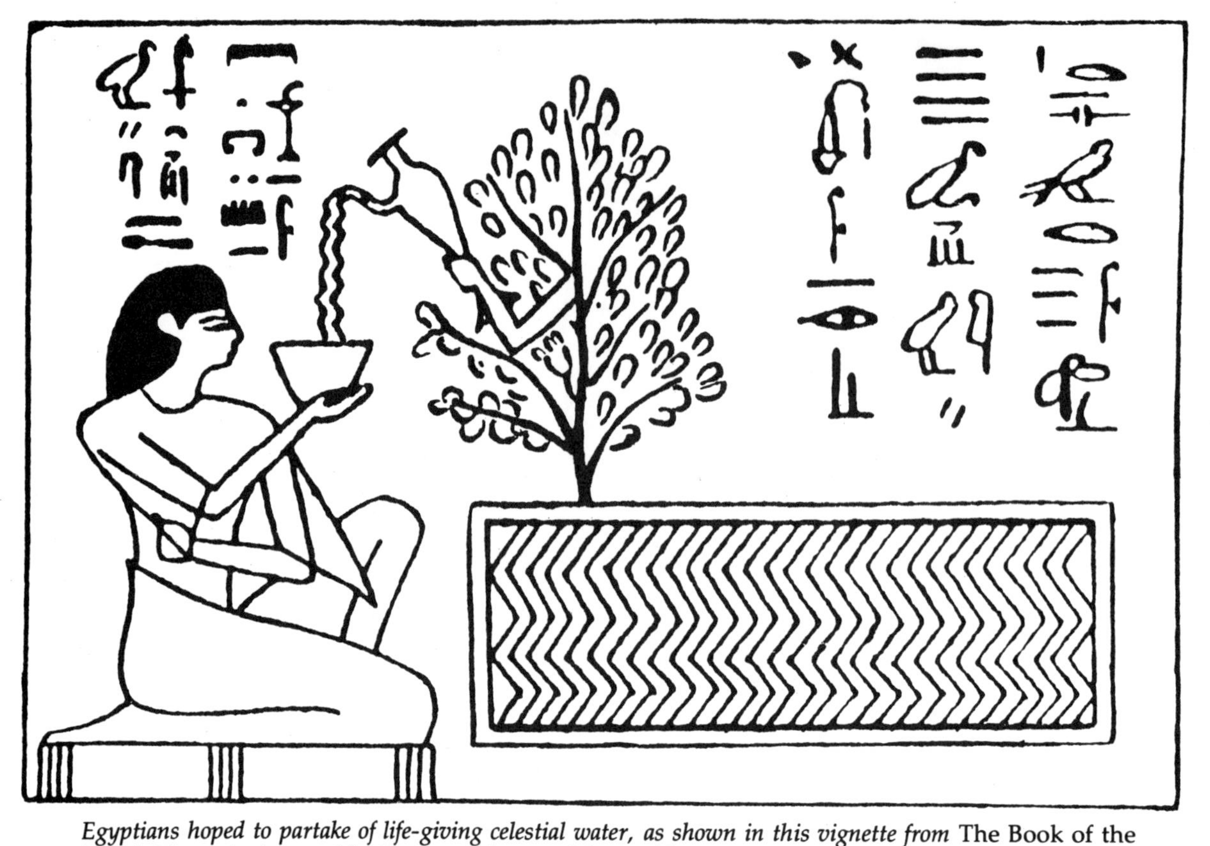

*Egyptians hoped to partake of life-giving celestial water, as shown in this vignette from* The Book of the Dead. *The water is served by the tree at the top of the world.*

## An Ancient Record of Out-of-the-Body Flight

There is, however, another possibility. Perhaps it was not the body of a dead king which was put into the sarcophagus, but a living man.

From a variety of records other than the Pyramid Texts Egyptologists have detected an ancient festival known as the 'Sed' or 'Heb Sed' Festival, also called the festival of the tail. James H. Breasted thought that this was probably 'the oldest religious feast of which any trace has been preserved in Egypt'.[23] Another eminent Egyptologist, Alexandre Moret, described it as follows:

> In most of the temples of Egypt, of all periods, pictures set forth for us the principal scenes of a solemn festival called festival of the tail, the Sed Festival. It consisted essentially in a representation of the ritual death of the king followed by his rebirth. In this case the king is identified with Osiris, the god who in historical times is the hero of the sacred drama of humanity, he who guides us through the three stages of life, death, and rebirth in the other world. Hence, clad in the funeral costume of Osiris, with the tight fitting garment clinging to him like a shroud, Pharaoh is conducted to the tomb; and from it he returns rejuvenated and reborn like Osiris emerging from the dead.[24]

For Moret, we note, this rejuvenation and return from death was purely symbolic: 'How was this fiction carried out?' he asks. We also note that this festival seems to have required the use of at least two specific garments worn by the king, one with a tail, and the tight-fitting robe of Osiris, in which Zoser is depicted in his *serdab*. A relief in a gallery under the Step Pyramid shows him wearing a kilt with a tail. The Sed Festival was celebrated upon the 30th anniversary of the king's reign, or so some say. Petrie found an inscription in the quarries of Hammamat referring to 'a Sed festival of Sirius' rising' in the second year of a king's reign. He thought that these festivals were then at fixed astronomical dates, and not dependent on the years of the reign.[25] The festival consisted of a number of rites, the significance of all of which is not yet fully understood. Dressed in his kilt with a tail, the king is depicted running or possibly dancing. In the case of Zoser, this event apparently took place in a special area called the Heb Sed Court within the walls of the large complex surrounding the Stepped Pyramid, several hundred yards northeast of the Pyramid of Unas.

Another architectural element involved in the festival seems to have been the curious boat pits found near certain pyramids, one of which is about 180 metres (200 yards) east of Unas' pyramid. From a tomb inscription at Thebes we know that during part of the festival one king was rowed around a lake in two special boats. One of these was the Solar or Day-boat and one the Night-boat. They seem to correspond to the boat pits facing east-west and north-south.

It is very interesting that these boats are also mentioned in the Pyramid Texts, but the general satisfaction with the tomb theory has prevented scholars from suggesting that the pyramids had anything to do with the Sed Festival. However, Selim Hassan noted that the inscription in the same Theban tomb describing the lake journeys makes it clear that these boats had to do with 'the ceremonial death and rebirth of the king, symbolized by his identification with the sun god, his journey in the Night-boat, and his return to this life in the Day-boat.'[26] This sounds much like the journey of Unas since, as we have heard, his journey involved an evening barge and a morning barge.

In fact, two further clues make it more than likely that what the Pyramid Texts actually represent is the core experience of this larger Heb Sed Festival. In the first place, the Hammamat quarry inscription recorded by Petrie describes the preparation of a royal sarcophagus in connection with a Heb Sed Festival. In the second, at one point the text of Unas states:

Unas opens the gate with the double doors,
Unas reaches the limit of the horizon
after Unas has put down there to the ground his robe
with the tail.[27]

*Relief from the Temple of King Sethos I at Abydos showing a king wearing a garment with a tail, part of the symbolism employed in the Heb Sed festival.*

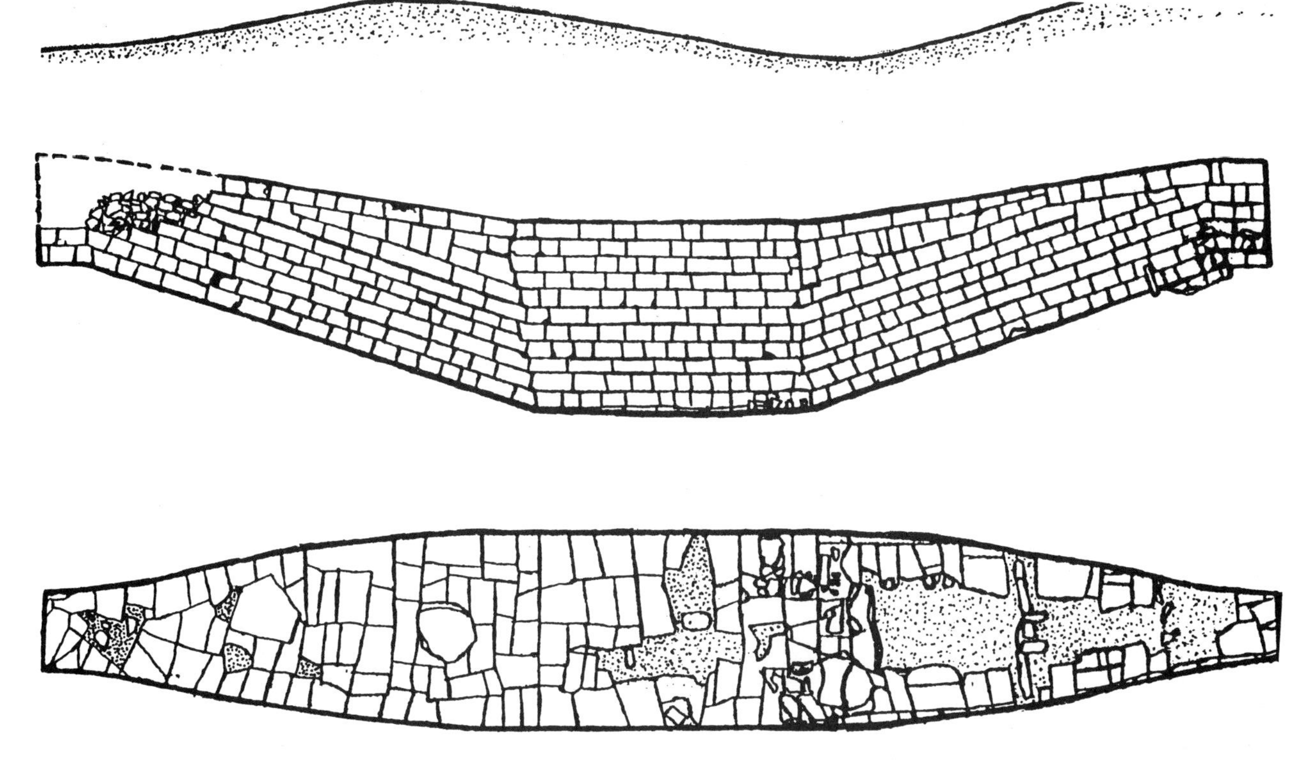

*The brick-built solar boat of Unas about 200 metres east of his pyramid was found under several metres of sand and debris. The boat is 36.5 metres (119.75 feet) long, 7.15 metres (23.45 feet) deep and 6.1 metres (20 feet) broad. It is orientated due east and even seems to have a rudder on its western (right-hand) end. Similar boats (or boat pits, as they are sometimes called) are scattered along the west bank of the Nile, usually within pyramid fields. Some are excavated out of bedrock; others, like that of Unas, are built of brick. One boat pit next to the Great Pyramid was uncovered in the early 1950s. It contained a large disassembled wooden boat which, when reconstructed, was too large to fit in the cavity. Most boat pits probably never contained wooden counterparts, but seem to have been the scenes of rituals connected with the Heb Sed festivals at the pyramids.*

*A brick built solar boat pit at Abusir with altars and platforms, showing that some boat pits were not intended to hold wooden boats.*

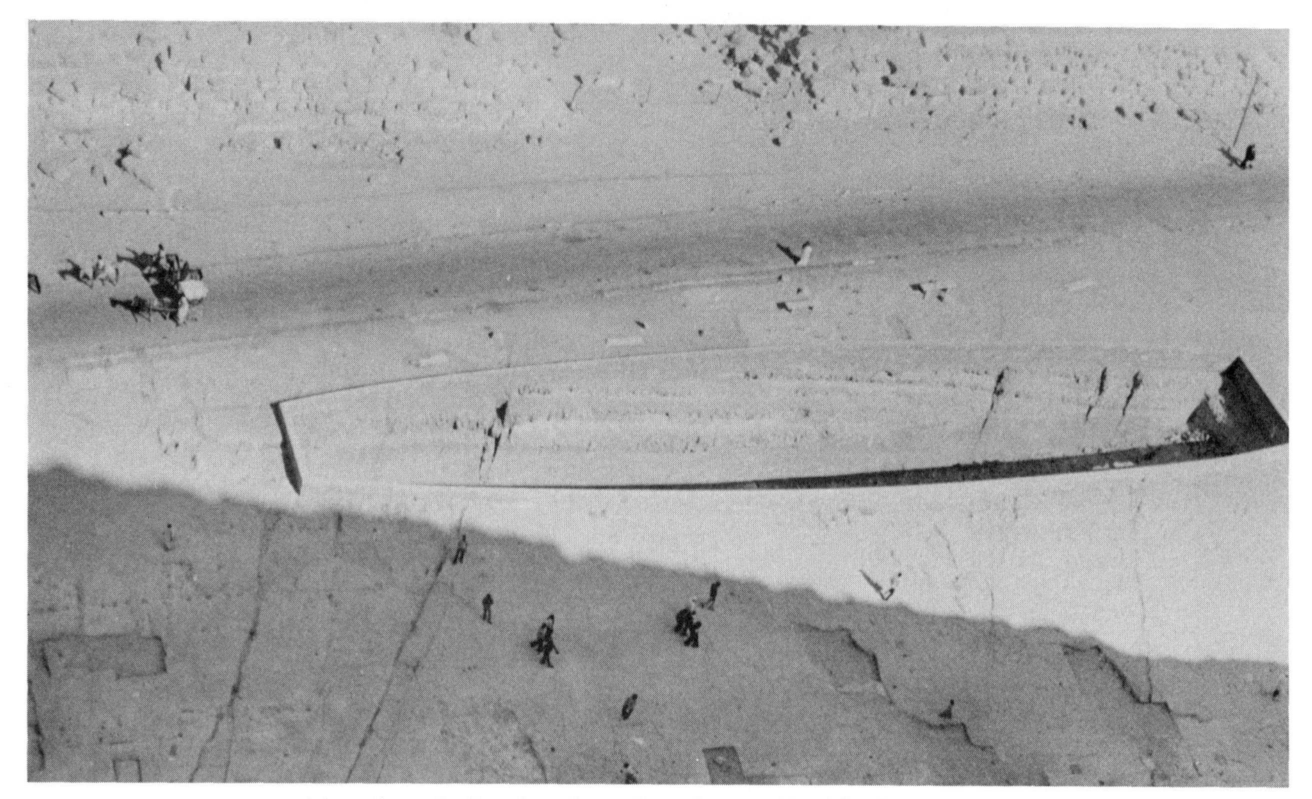

*A large boat pit aligned north-south on the east side of the Great Pyramid.*

*The spirit of the deceased stands between two celestial boats. Detail from the Papyrus of Nu.*

A major piece of the puzzle seems to fall into place. What can this be but the tailed garment of the Heb Sed? This tail probably symbolized man's animal body which must be left behind on the king's journey. But what kind of journey could this have been from which the king returned alive and restored, and yet took place inside a pyramid?

The astronomic imagery of the texts leads us to suggest that with its lid in place the sarcophagus functioned as an initiation chamber and that the 'funeral ritual' was a series of hypnotic formulas repeated to the subject to

*A gate in the Temple of Karnak with inscriptions pertaining to the Heb Sed festival is flanked by four images of a king carrying the* was *sceptre and the* ankh. *He wears a kilt to which a tail is attached.*

induce what we today call an out-of-body experience. If Unas indeed left his body, adventured into higher realms and then returned to his body, would he not awake and arise as one who had literally ascended to heaven and seen the gods?

In the Unas text there are a number of statements which supposedly pertain to the soul of the king in the afterlife, but which make much more sense if they are applied to a living king in his physical body after his initiation. On the same east wall through which he left, the texts record what sounds very much like the return of his spirit as Unas was called back to his body and reawakened to life on earth. In fact, he is specifically told: 'Put on thy body.'

O Unas , stand up. . . . [28]
O Unas, come up to me, betake thee to me. . . . [29]
O Unas, stand up, receive this thy bread from my hand![30]
Awake, Unas! Turn around, Unas![31]
How changed, how changed (is thy state)!
(Therefore) protect thy children!
Beware of thy border which is in Earth!
Put on thy body and come towards them![32]

If Unas returned from an out-of-body flight, he may have been gone only a few hours but his several risings and settings suggest a longer interval of three or more days. After being out of the body for even a few hours, much less three days, possibly he might at first find it cold, stiff and uncomfortable. It is appropriate, then, to find what sounds like descriptions of exercises to stir up the circulation: 'Stir up . . . turn around, O! Unas, stand up . . . '[33] and later, 'Osiris Unas, dance!'[34]

After the calisthenics there follows a great feast and celebration. Unas is offered beer, wine, bread, delicate meats and an incredible amount of food, oil and produce of every kind: 'Sit down to the thousand of bread, to the thousand of beer! The roast of thy double rib is from the (king's) slaughter house.'[35] This feast is usually seen by Egyptologists as offerings made to the *ka* or 'double' of Unas, another element of man's total being which the Egyptians believed remained behind on earth after the departure of the higher soul.[36] However, during the course of the feast an offering is made to Unas which he is not to consume, and the text reads: 'Prevent him from swallowing it!'[37] — a concern which is appropriate only with a physical body. We are told, indeed, that 'Unas is he who went and came back'.[38]

If the Text of Unas were an isolated example, it might more easily be disregarded. But other ancient texts in the pyramids of Tety, Pepy I and

Pepy II, also at Saqqara, give highly parallel descriptions of similar journeys, including statements that Tety or Pepy has not died: 'He does not give himself to Osiris. Tety did not die a death.'[39]

The bona fide royal tombs of the Egyptian monarchs differ entirely from the pyramids in style; typically they are chambers cut into rock hills, as in the Valley of the Kings across the Nile from ancient Thebes, and they actually look like tombs. There are no texts in these tombs indicating that the soul has not been given over to Osiris or that the king returned to his body. Nor do we find these indications in any of the many texts collectively known as the *Book of the Dead*. On the contrary, the *Book of the Dead* is largely concerned with the very judgment and 'weighing of the heart' from which Unas was pointedly exempted. Appropriately, the best evidence that the pyramids were not tombs comes from the pyramid texts themselves, and the interpretation that these inner chambers were the scenes of initiations and not burials perfectly accounts for all the empty pyramid sarcophagi and the fact that no original burial has ever been found in any pyramid in Egypt.

Obviously, the greatest single question that emerges from all this is whether Unas and the others actually did what the pyramid texts appear to claim they did, or whether it was merely an empty ritual wherein everything that was said to happen took place only in imagination. Initiation as an explanation of what took place inside the pyramids has been put forward many times before,[40] but invariably by esoteric writers relying on intuition and not on documented evidence.

As it happens, astonishing new evidence far from the field of Egyptology has emerged in just the last few years which documents that others have made journeys like Unas, and allows us to see these ancient texts in a new light. The 20th century is catching up with the magic of the pharaohs — and as the doors to the mysteries of the past swing open, we find ourselves in a world fantastic beyond belief.

CHAPTER SIX

# ON THE TRAIL OF UNAS

## Modern Remote Viewing Experiments

Stanford Research Institute, Menlo Park, California, 27 April 1973:

6:03:25 PM There's a planet with stripes.
6:04:13 PM I hope it's Jupiter.
I think that it must have an extremely large hydrogen mantle. If a space probe made contact with that, it would be maybe 80,000 to 120,000 miles out from the planet surface.
6:06 PM So I'm approaching it on the tangent where I can see it's a half moon; in other words, half lit, half dark. If I move around to the lit side, it's distinctly yellow to the right.
6:06:20 PM Very high in the atmosphere there are crystals — they glitter — maybe the stripes are like bands of crystals, maybe like the rings of Saturn, though not far out like that; very close within the atmosphere. I bet you they'll reflect radio probes. . . .
6:08 PM Now I'll go down through. It feels really good there (laughs). Inside those cloud layers, those crystal layers, they look beautiful from the outside — from the inside they look like roiling gas clouds — eerie yellow light, rainbows.[1]

This is a transcript recording the report of artist Ingo Swann as he perceived the planet Jupiter during a remote viewing experiment. At the time Swann was in an armchair at Stanford Research Institute. He was monitored by SRI physicists Russell Targ and Harold Putoff who describe this and related experiments in their book, *Mind-Reach*. Swann was joined in his attempt by Harold Sherman, then in Arkansas, who had earlier documented his ability to communicate via extrasensory perception.

Under careful observation by Targ and Putoff, Swann and Sherman attempted their remote viewing of Jupiter before the approaching flyby of Pioneer 10 in the hope they might report something that would be later verified by the space probe. The two men produced highly parallel descriptions of the giant planet. While their accounts were not at odds with anything that was already known about Jupiter nor with the additional information radioed back by Pioneer 10, the NASA probe was not instrumented to verify their highly visual descriptions.

On 11 March 1974 Swann and Sherman tried again. This time they targeted on the planet Mercury before the flyby of another space probe, this one called Mariner 10. Less was known of Mercury than of Jupiter beforehand. The two men described a weak magnetic field, a thin atmosphere and a 'tail' of helium streaming out from Mercury away from the sun. None of these features was then known or predicted by astronomers; all of them were verified by Mariner 10.[2]

Of course, we are dealing with aspects of man and nature here which cannot be accounted for within the beliefs of many individuals, be they laymen or scientists. One reviewer of Targ and Putoff's work commented: 'This is the kind of thing that I would not believe in even if it existed.'[3] The evidence that it does exist, however, is now overwhelming. Among those who have testified on the subject is Carl Gustav Jung. In *Memories, Dreams, Reflections* the great psychologist described his own experience, amazingly similar to those of Unas, Swann and Sherman. Jung was 68. He was the author of hundreds of books and papers ranging from psychiatric studies to religion and alchemy. Behind him was a great career dealing with all aspects of psychology; yet even Jung found his experience 'extremely strange'.

> At the beginning of 1944 I broke my foot, and this misadventure was followed by a heart attack. In a state of unconsciousness I experienced deliriums and visions which must have begun when I hung on the edge of death and was being given oxygen and camphor injections. The images were so tremendous that I myself concluded that I was close to death. My nurse afterward told me: "It was as if you were surrounded by a bright glow." That was a phenomenon she had sometimes observed in the dying, she added. I had reached the outermost limit, and do not know whether I was in a dream or an ecstasy. At any rate, extremely strange things began to happen to me.
>
> It seemed to me that I was high up in space. Far below I saw the globe of the earth, bathed in a gloriously blue light. I saw the deep blue sea and the continents.

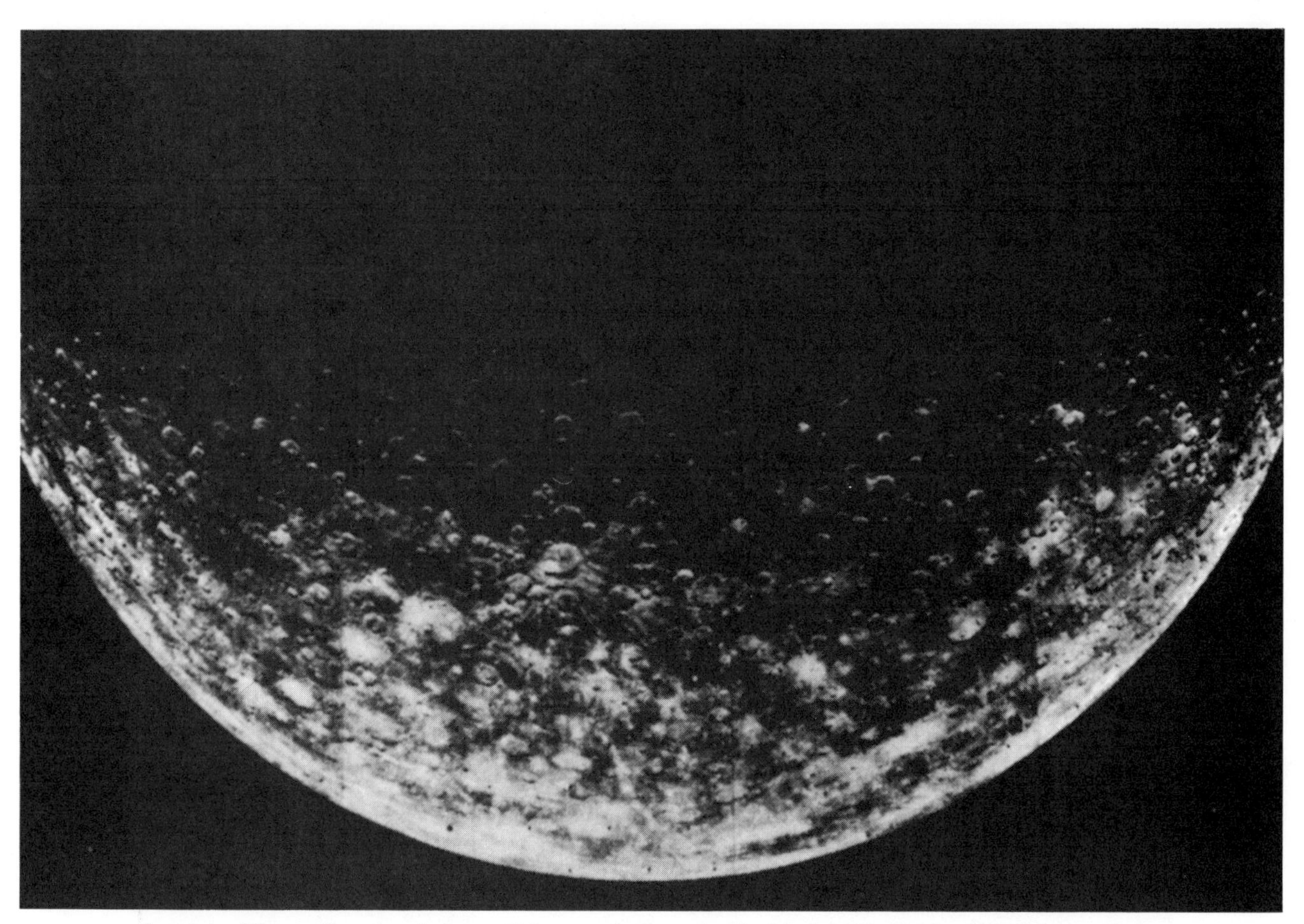

*A mosaic of the southern hemisphere of Mercury photographed by Mariner 10, 29 March 1974 from a distance of 200,000 kms (125,000 miles) during the mission which corroborated descriptions of the planet produced by Ingo Swann and Harold Sherman in remote viewing experiments.*

> Far below my feet lay Ceylon, and in the distance ahead of me the subcontinent of India. My field of vision did not include the whole earth, but its global shape was plainly distinguishable and its outlines shone with a silvery gleam through that wonderful blue light. In many places the globe seemed coloured, or spotted dark green like oxydized silver. Far away to the left lay a broad expanse—the reddish-yellow desert of Arabia; it was as though the silver of the earth had there assumed a reddish-gold hue. Then came the Red Sea, and far, far back—as if in the upper left of a map—I could just make out a bit of the Mediterranean. My gaze was directed chiefly toward that. Everything else appeared indistinct. I could also see the snow-covered Himalayas, but in that direction it was foggy or cloudy. I did not look to the right at all. I knew that I was on the point of departing from the earth.
>
> Later I discovered how high in space one would have to be to have so extensive a view—approximately a thousand miles! The sight of the earth from this height was the most glorious thing I had ever seen.[4]

The globe of the earth was not all that Jung saw while in space, but before he could explore this realm his doctor quite literally called him back to life. Jung was so intrigued with this experience that he found it 'extremely disappointing' to have to return to life on earth. But return he did, and

*The earth seen from space. The Horn of Africa and the Arabian Peninsula are seen at the top.*

lived another 17 years. Jung confessed that he 'would never have imagined that any such experience was possible'.

We live in an age when the greatest psychologist of the century had trouble even imagining such things until it happened to him. But times change. Astronauts who participated in manned space flights have since reached altitudes much greater than Jung's and report that even when seen in-the-body the beautiful blue sphere we call earth is no less dramatic than in the psychologist's account. In the decades since Jung's flight over the Indian Ocean, the examination of psychic and paranormal experiences which in earlier days could only be characterized as mystical has recently moved from a fringe activity into the laboratory. The SRI experiments with Swann and Sherman were part of a series of investigations conducted by laser physicists Targ and Putoff who were initially exploring 'quantum biological effects'.[5] Their results with Swann (and with Uri Geller some time earlier) led them to do remote viewing tests with a wide range of individuals. Under the most stringent controls that can be devised, positive evidence documenting the reality of remote viewing, out-of-body and related experiences is now a matter of fact.

In other experiments Targ and Putoff gave Swann a set of geographical coordinates—locations described only in terms of latitude and longitude—and asked him to report what he found at those places. Under controlled conditions Swann accurately described widely separated points on the earth's surface in seven attempts out of ten. The results of two attempts were neutral or ambiguous; only in one case was there a clear lack of correspondence.

These experiments were so tightly controlled that Targ and Putoff themselves did not know until the last minute what coordinates Swann would be asked to describe. To avoid the possibility of 'subliminal cueing', the targets were chosen by colleagues out of the building who telephoned the coordinates to Targ and Putoff at the beginning of each session. Nor can Swann's results be attributed to memory. He accurately described buildings and other details not shown on maps on an island in the south Indian Ocean. A freehand map he drew of this place, Kerguelen Island, while viewing it remotely brings to mind the maps of the ancient sea kings studied by Hapgood, and points out the useful applications to which abilities like his could have been put in antiquity. Although fairly crude, Swann's map of Kerguelen and accompanying comments contain sufficient detail of the coast to be valuable to a navigator looking for a port. This map was a first attempt by an untrained individual with natural psychic ability.

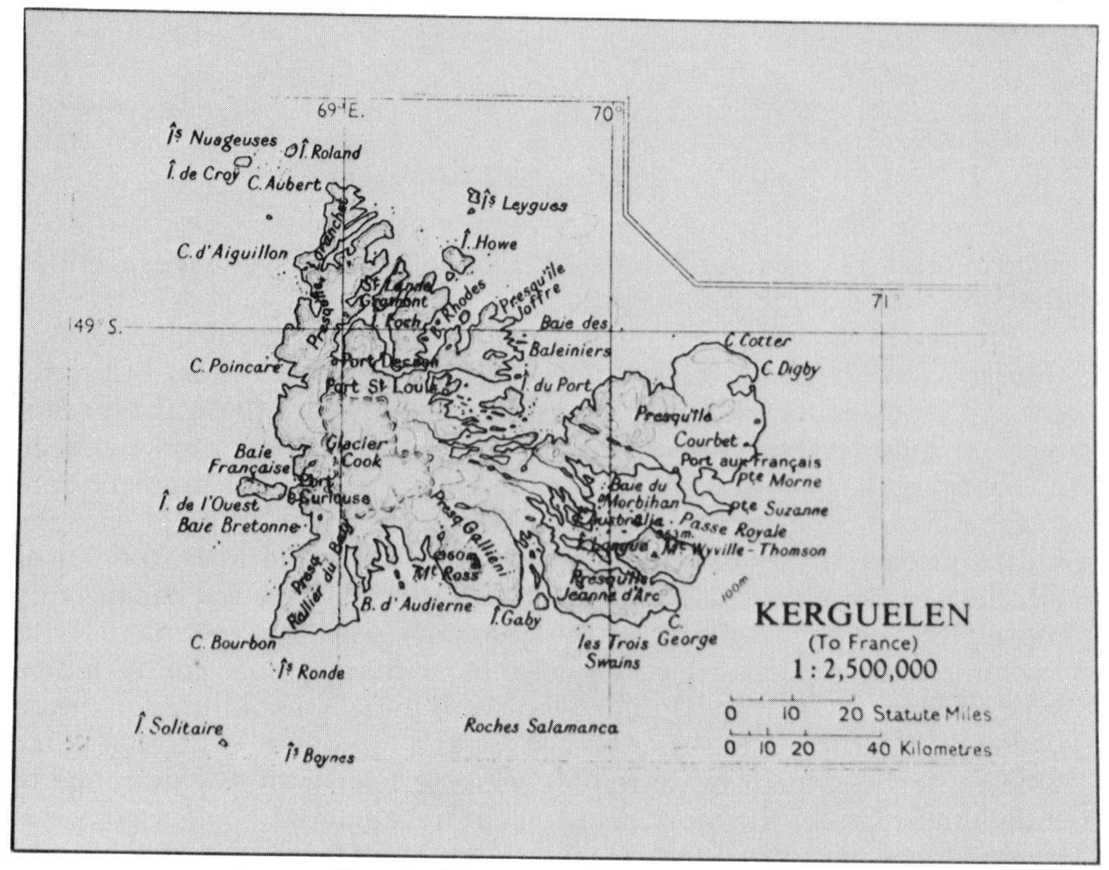


*Kerguelen Island in the South Indian Ocean.*

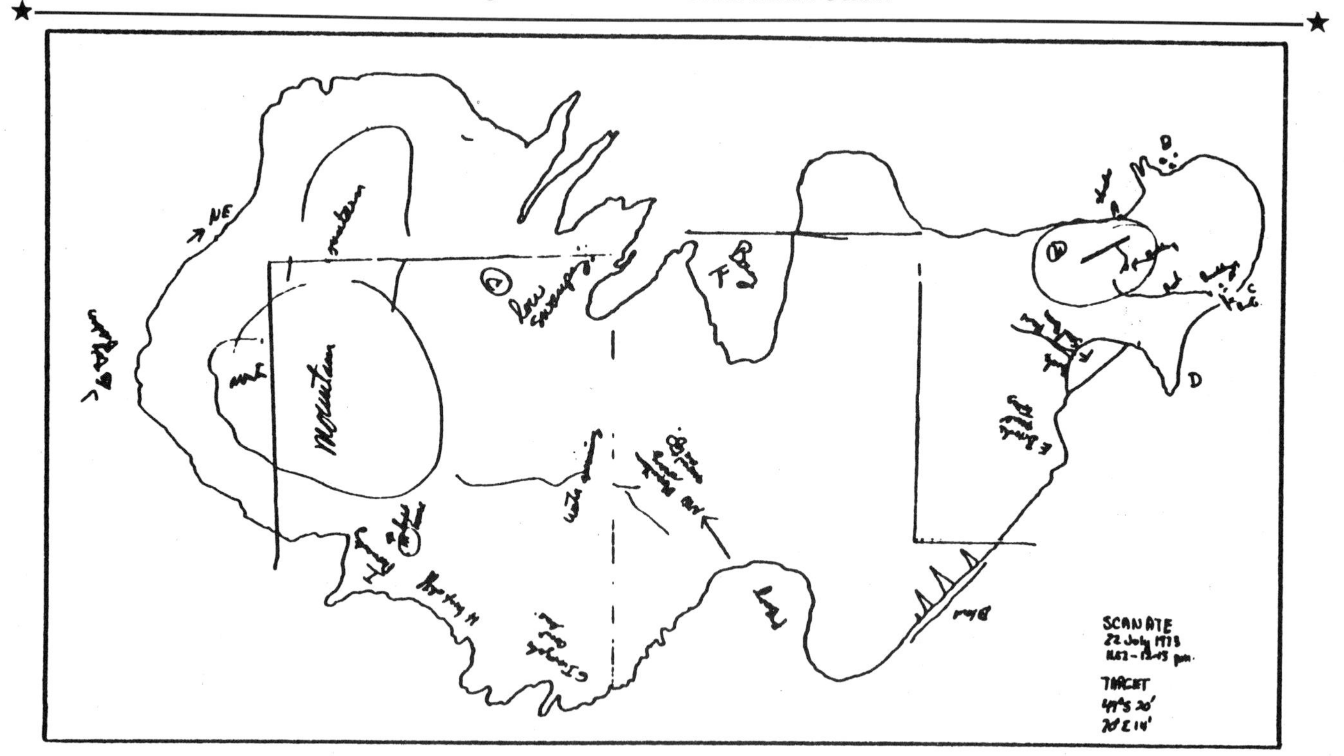


*A freehand map of Kerguelen Island drawn by Ingo Swann during a remote viewing experiment at Stanford Research Institute in 1973.*

## The Mystical Experience of Cosmic Travel

Of course, it is one thing to map a 50 mile long island and another to map a continent such as Antarctica. But if Swann could achieve what he did on a first attempt, what could ancient priests have done who devoted lifetimes to these activities? They certainly could have provided a 'distant early warning system' reporting any troop movements within hundreds of miles of their borders. The ancient Akbar Ezzemann manuscript reports that Egypt was long unconquerable because of the 'magic' possessed by the Egyptians. Even with superior numbers and armaments, invaders would be at a distinct disadvantage if the pharaoh and his generals were continually informed of their positions.

Remote viewing is now a documented fact, but was this the same activity in which Unas was engaged? The answer is both yes and no. If we have correctly interpreted what took place beneath the sands of Saqqara over 40 centuries ago, Unas underwent what we today call an out-of-body experience. Remote viewing is a neutral term covering a spectrum from simple, conscious clairvoyance to profound and sometimes lengthy separations of consciousness (and certain elements of being) from the physical body. Ingo Swann produced accurate remote viewings while fully conscious, drinking coffee and smoking a cigar, which led those who monitored him to suggest that his was a function of clairvoyance, rather than out-of-body flight.[6]

With minimal physical resources, it is not surprising to find that many archaic peoples made use of clairvoyance in coping with their environment. The Selknam of Tierra del Fuego described this power as 'an eye which stretches out of the magician's body . . . in a straight line towards the object that it has to observe while still remaining united with the magician.'[7]

For the person experiencing it, such clairvoyance is the least involving of the many shades of altered states of consciousness. Even in this case there appears to be interaction on some physical level between the subject and the object or creature viewed. Experiments with animals that were 'viewed at a distance' as well as with super-sensitive electronic equipment indicate both animals and machines can in some way sense, register or detect the presence of the viewer at the time they are being observed.[8]

Out-of-body experiences, on the other hand, are among the most involving paranormal states, and the experiential intensity of the interaction with distant realms and beings is much greater. In *Journeys Out of the*

*Body,* Robert Monroe gives detailed descriptions of such experiences launched from sleep. Instead of merely directing his attention toward the distant object to be viewed, as Swann does, Monroe seems to take his perceptions, emotions and sexuality with him as he leaves his physical body and travels *in* a subtle body.

There are even documented cases in which a person having this kind of experience was perceived by fully conscious observers at the site to which he or she journeyed! In October 1863 S. R. Wilmot of Bridgeport, Connecticut was on board a steamer heading from Liverpool to New York. On a night following a nine day storm he dreamt that his wife came to the door of his stateroom, clad in her nightdress. 'At the door she seemed to discover that I was not the only occupant of the room, hesitated a little, then advanced to my side, stooped down and kissed me, and after gently caressing me for a few moments, quietly withdrew.' His stateroom companion, who was lying wide awake in the berth above Wilmot, saw exactly what Wilmot had seen in his dream.

Returning to Connecticut, Wilmot learned that on the night in question his wife had lain awake concerned about the weather and thinking of him. At about four o'clock in the morning it seemed to her that she went out to seek him. Crossing the stormy sea, she came at last to his ship and found his stateroom. 'Tell me,' said she, 'do they ever have staterooms like the one I saw, where the upper berth extends further back than the under one? A man was in the upper berth, looking right at me, and for a moment I was afraid to go in, but soon I went up to the side of your berth, bent down and kissed you, and embraced you, and then went away.' The description given by Mrs. Wilmot of the steamship was correct in all particulars, though she had never seen it in the flesh. At the time of her visit the ship was more than a thousand miles out at sea.[9]

Naturally, Mrs. Wilmot's escapade is beyond laboratory verification, but some out-of-body experiences have been measured and documented under highly controlled conditions. Parapsychologist Charles Tart of the University of California at Davis monitored a sensitive identified as Miss Z who attempted to discover a concealed five digit number while having an out-of-body experience. Miss Z did so against odds of 100,000 to one. She was hooked up with wires attached to machines that recorded her brain-wave patterns, rapid eye movement, basal skin resistance, galvanic skin resistance, heart rate and blood volume. There was a clock in the same room, and on the night she read the number, she reported the time as between 05.50 and 06.00. At 05.57 her brain-wave readouts showed strange patterns noted on previous occasions when she claimed to be out of

*An out-of-body experience depicted by the Chinese artist T'ang Yin (1470–1523) in his painting 'Dreaming of Immortality in a Thatched Cottage'.*

the body.[10]

The trail of Unas leads to recognition of the fact that under a variety of circumstances some conscious sentient part of a human being normally resident in the physical body can temporarily leave it, journey far or not at all, return to residence in the body again, remember and report its experience. Of course, only the tiniest fraction of such experiences happen within a laboratory. There are undoubtedly many people who 'would not believe in this kind of thing even if it existed'; denial, however, does not render immunity from an out-of-body experience, and there is nothing like the experience itself to change one's mind.

Not forgetting contemporary lab work, the best evidence that these experiences are real is that they happen to a great many people. Over thousands of years people from various cultures throughout the world have reported these flights from the body. Usually, they seem to happen spontaneously at times of extreme physical crisis such as being injured in an accident, wounded in battle, falling from a height, or at times of grave illness, while under anaesthesia during operations and near, at and beyond the point of medical death.

There are countless reports of injured people who 'wake up' to the realization that they are viewing their own disabled body, sense they are free to travel, visit a close relative or friend, and return to their body often after the worst is over and the body helped. The visited relative or friend sometimes reports having seen, felt, sensed or thought of that person at the time of the incident.

Even an experience of this type, in which the travellers recount what largely seems to be the customary physical world, usually has a profound effect upon the individual concerned. For many, it takes on a religious importance and amounts to the discovery or rediscovery of their souls, altering the person's perspective in a deep and lasting way. Wounded in action in Italy during World War I, a youthful Ernest Hemingway had an out-of-body experience which he incorporated in *A Farewell to Arms*. This experience seems to have been the source of the famous Hemingway predilection to seek confrontations with death. Much of his writing is pervaded with the same fascination.

## The Celestial Visions of a Prophet

Not all out-of-body travellers experience only views of the physical earth. Some perceive beings and places profoundly removed from the customary world. These visions added to an out-of-body flight are almost invariable features in the careers of prophets, saints and religious reformers. Among the most powerful accounts of this kind of experience in recent times is that of the Oglala Sioux medicine man, Black Elk. When he was a boy of nine in the latter part of the 19th century, Black Elk had a transforming vision that ruled his entire life. Typically, it began with an illness. The experience is described in *Black Elk Speaks*.

> There was a man by the name of Man Hip who liked me and asked me to eat with him in his tepee.
>
> While I was eating, a voice came to me and said: "It is time; now they are calling you." The voice was so loud and clear that I believed it, and I thought I would just go where it wanted me to go. As I came out of the tepee, both my thighs began to hurt me, and suddenly it was like waking from a dream, and there wasn't any voice.

The next day his condition worsened. He could not walk and his legs, arms and face became badly swollen.

> When we had camped again, I was lying in our tepee and my mother and father were sitting beside me. I could see out through the opening, and there two men were coming from the clouds, headfirst like arrows slanting down, and I knew they were the same that I had seen before. Each now carried a long spear, and from the points of these a jagged lightning flashed. They came clear down to the ground this time and stood a little way off and looked at me and said: "Hurry! Come! Your grandfathers are calling you!"

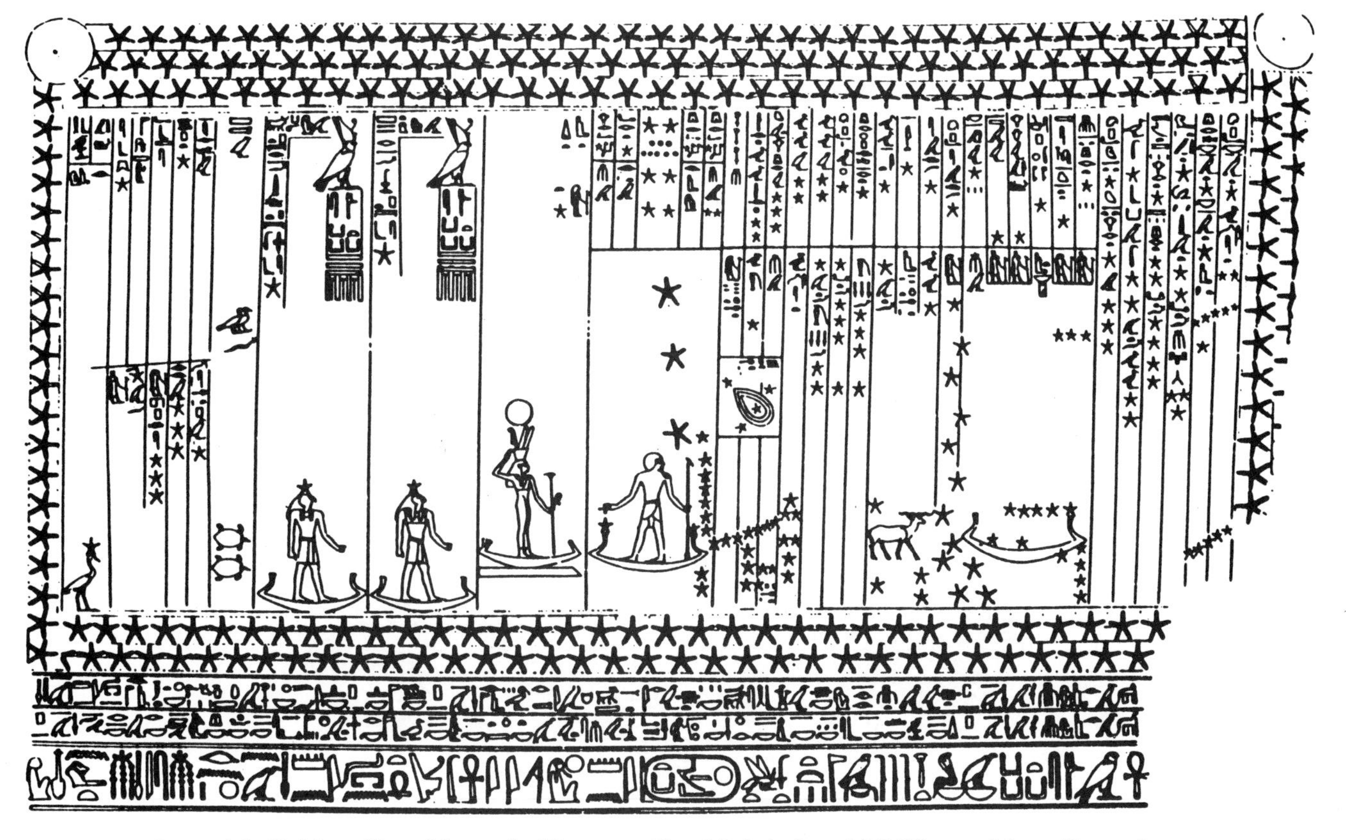

*The south half of the ceiling of the tomb of Senmut at Der el Bahri, about 1473 BC, one of the earliest and most important of the major Egyptian star maps. The extreme left hand column with a bird and star at the bottom is Venus; the next column describes Mercury. The first barque from the left is Saturn; the next is Jupiter, then Isis as an attendant deity immediately to the left of Orion, easily identified by the three large stars of the Belt of Orion. The rest of the map is largely taken up by descriptions of the decans, stars which define ten degree segments of the night sky.*

*An ornamental Tree of Life incorporating the three cosmic zones. From an ancient Assyrian wall carving.*

*A bamboo decoration of the Or-Danom Dayak people of Borneo shows the world divided into three cosmic zones. In the Underworld the cosmic tree grows from a serpent, the axis of which continues in the middle world of ordinary life as the centre pole of the house. Above the roof of the house the axis continues, surmounted by a bird.*

Then they turned and left the ground like arrows slanting upward from the bow. When I got up to follow, my legs did not hurt me any more and I was very light. I went outside the tepee, and yonder where the men with flaming spears were going, a little cloud was coming very fast. It came and stooped and took me and turned back to where it came from, flying fast. And when I looked down I could see my mother and my father yonder, and I felt sorry to be leaving them. . . .

Then there was nothing but the air and the swiftness of the little cloud that bore me and those two men still leading up to where white clouds were piled like mountains on a wide blue plain, and in them thunder beings lived and leaped and flashed. . . .

Then as we walked, there was a heaped up cloud ahead that changed into a tepee, and a rainbow was the open door of it; and through the door I saw six old men sitting in a row.

The two men with the spears now stood beside me, one on either hand. . . . And the oldest of the Grandfathers spoke with a kind voice and said: "Come right in and do not fear." . . . So I went in and stood before the six, and they looked older than men can ever be — old like hills, like stars.

The six Grandfathers came from the four directions, the sky and the earth. They are identified as 'the powers of the World'. Then the first Grandfather spoke again:

Behold them yonder where the sun goes down, the thunder beings! You shall see, and have from them my power; and they shall take you to the high and lonely centre of the earth that you may see; even to the place where the sun continually shines, they shall take you there to understand.

The first Grandfather then gave Black Elk a wooden cup full of water, 'and in the water was the sky'. This was 'the power to make live'. The Grandfather also gave him a bow — 'the power to destroy'. The other Grandfathers also gave Black Elk their powers. The last one told him: 'My boy, have courage, for my power shall be yours, and you shall need it, for your nation on the earth will have great troubles.' Black Elk received the emblems and viewed the symbols of power and travelled the heavens with visions of thunder beings, animal spirits and virgins. And then, at 'the centre' of the world:

I was standing on the highest mountain of them all, and round about beneath me was the whole hoop of the world. And while I stood there I saw more than I can tell and I understand more than I saw; for I was seeing in a sacred manner the shapes of all things in the spirit, and the shape of all shapes as they must live together like one being. And I saw that the sacred hoop of my people was one of many hoops that made one circle, wide as daylight and as starlight, and in the centre grew one mighty flowering tree to shelter all the children of one mother and one father. And I saw that it was holy.

Finally, it was time for Black Elk to return to the body. A spotted eagle was sent with him to protect him.

I was all alone on a broad plain now with my feet upon the earth, alone but for the spotted eagle guarding me. I could see my people's village far ahead, and I walked very fast, for I was homesick now. Then I saw my own tepee, and inside I saw my mother and my father bending over a sick boy that was myself. And as I entered the tepee, someone was saying: "The boy is coming to; you had better give him some water." Then I was sitting up; and I was sad because my mother and my father didn't seem to know I had been so far away.

Like countless prophets and spiritual ministers before and since, Black Elk's vision not only equipped him to be a healer of his people, but continued to serve then and now as a guiding vision for his community in its trials and struggles. The laboratory tests and other accounts of out-of-body experiences we have surveyed are only part of the tip of an enormous iceberg; hundreds or thousands of examples could be found. Contemporary laboratory scientists in this area are in fact only catching up with many earlier practitioners, from Jung and Black Elk to Unas. Yet this research enables those who have never had (or do not recall) the experience to be aware of it and perhaps to regard our explorations more seriously. It also enables us to throw new light on the story of Unas and to see many related phenomena from a fresh perspective. But if we are to catch up with Unas, we must know more than that some part of him flew above the earth and deep into space. What was the purpose of his journey? What was the nature of his star ship?

CHAPTER SEVEN

# INITIATION

## The Universal Ritual of the Magical Flight

We have, it seems, stumbled upon a great secret. In viewing the experience of Unas from the perspective of an out-of-body flight, the remarkable thing is not only that this interpretation makes such good sense of the texts (and physical remains) in his pyramid, nor even that contemporary laboratory science is now able to confirm that such flights actually take place. The really extraordinary result of this perspective is that it reveals the central initiatory experience behind most of the great religions, mysteries and cosmologies of the ancient world.

The flight of Unas intersects first of all with shamanism. According to Mircea Eliade, the well-known historian of religion, 'shamanism in the strict sense is preeminently a religious phenomenon of Siberia and Central Asia'.[1] Throughout that immense area, the shaman was the centre of the religious and magical life of the community. Yet he was something more than a priest, soothsayer, sorcerer, witch doctor or magician. According to Eliade, 'the shaman specializes in a trance during which his soul is believed to leave his body and to ascend to the sky or descend to the underworld'.[2]

He was thus partly magician and, since he had healing powers, something of a medicine man; but not every magician or medicine man was a shaman. Only the shaman had mastered the technique of the 'magical flight'. While the details of the rituals surrounding the shaman varied from culture to culture, a specific figure who undertook these flights was found among archaic peoples in North and South America, Australia, Africa and

Europe, as well as in all parts of Asia and the Pacific Islands — literally all over the earth.

An individual attained the status of shaman either by having a spontaneous experience of such a flight, like Black Elk, or through an initiation, a ritual in which the flight was induced. These so called mystical or magical flights were greatly prized by archaic peoples. Not only in Central and Northern Asia, but virtually everywhere, the individual who experienced these flights was the hub of the community's religious life. Nevertheless, he was not a sacrificing priest and was not otherwise required to perform the functions of a priest — although he might be called upon for assistance in special cases.

These magical flights greatly puzzled early ethnologists and anthropologists; on the one hand they were so widely and literally reported, and on the other they were outside the belief-structures of the investigators. Like Jung, they never would have thought that any such experience was literally possible. Considering the core experience of the most important individual within these cultures as so much fantasy, it is little wonder that the modern world developed thoroughly deprecatory opinions of ancient and archaic peoples.

However Eliade, who spent a lifetime studying shamanism and ancient cosmologies, concluded that these flights were not fantasies. He emphasizes that the flight of the shaman was 'a flight in the strictest sense of the word'.[3] In the definitive work, *Shamanism*, Eliade describes how archaic peoples the world over divided the universe into three cosmic zones — sky, earth and underworld. 'The preeminently shamanic technique is the passage from one cosmic region to another — from the earth to the sky or from earth to the underworld. The shaman knows the mystery of the breakthrough in plane.'[4]

Unas was far from one of a kind. The destination of the shamans, too, was the circumpolar stars, and the pole star in particular. In the mythic imagery of many Asian and Siberian peoples, the sky is imagined as a tent in which the stars are holes for light or windows of the world.[5] The pole star is said to keep the tent in place, and is referred to as the Nail Star or Sky Nail. And so also derives the image of the pole star as a stake supporting the tent of heaven. Among certain groups of Tartars, the stars are pictured as a herd of horses and the pole star 'is the stake to which they are tethered'.[6] Hence other peoples conceive the pole star as a pillar — an Iron, Golden or Solar pillar. It is the same basic image as the Cosmic Churn of Horus and Set, the column of light in Plato, the central pillar of the world represented as the World Tree, the Central Mountain, the Ladder or Pole of

*Re-vivification by solar rays: a scene from the Papyrus of Pa-di-Amon (21st dynasty, 1113-949 BC) shows a mummy-like figure being re-vivified by rays of stars and suns pouring from the head of Horus which is beneath the vault of heaven. Above is a Djed column with the face of Osiris, flanked by Isis and Nephthys and two symbols for the West.*

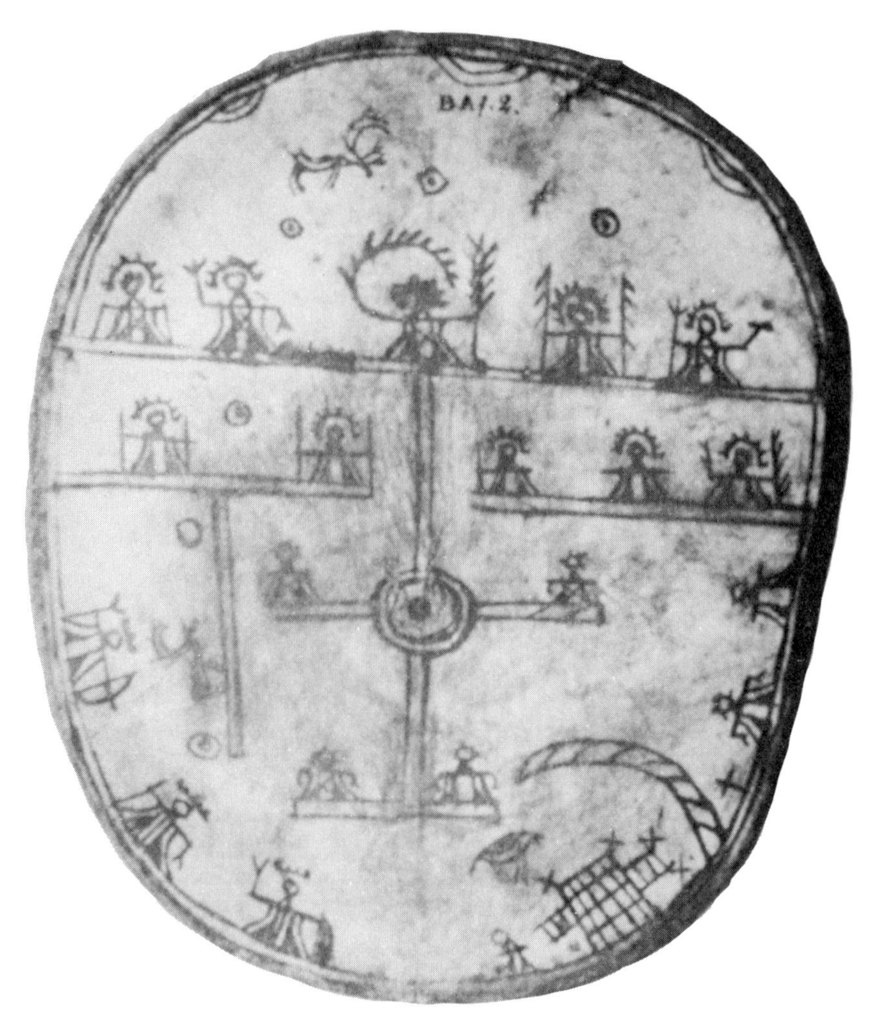

*A shaman's drum from Lapland, showing the world's centre with the* axis mundi *uniting the three worlds.*

other cultures that all reduce to the World Axis that unites the three cosmic zones. Whether it was called the Golden Pillar or the Sky Nail, the pole star was also seen as an opening: 'It is through this hole that the gods descend to earth . . . it is through the same hole that the soul of the shaman in ecstasy can fly up or down in the course of his celestial . . . journeys.'[7]

Due to their power of flight, souls or spirits are often represented as birds. Long before Jonathan Livingston Seagull, Siberian shamans knew that the bird that flies the highest, sees the farthest. According to the Tungus shamans:

> Up above there is a certain tree where the souls of the shamans are reared, before they attain their powers. And on the boughs of this tree are nests in which the souls lie and are attached. The name of the tree is Turru. The higher the nest in this tree, the stronger will the shaman be who is raised in it, the more will he know, and the farther will he see.[8]

It is worth emphasizing, as Eliade does, that this cosmology of the three cosmic zones united by a world axis is one of the truly universal patterns of

*The Scandinavian World Tree, Yggdrasil, showing the three cosmic zones and the tree as axis passing through the centre of the earth. It was from Yggdrasil that Odin, wounded by his own spear, hung himself for nine days and nine nights before attaining rejuvenation.*

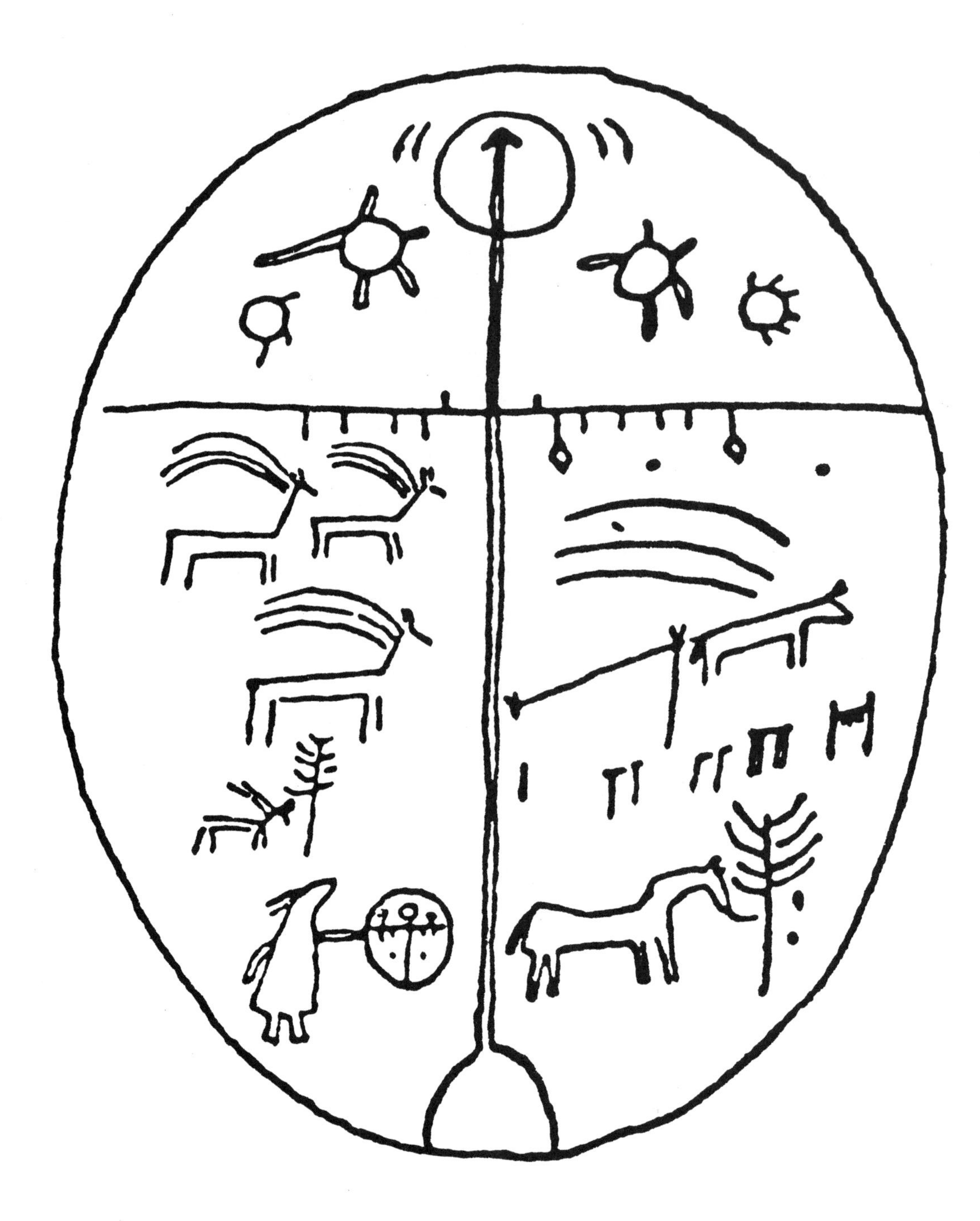

*A painting on a shaman's drum with an axis pointing upward to a hole among the stars in the upper world. Below left is the shaman holding his drum and gazing upward.*

*A serpent coils up the World Tree between its two fruits. The Tree is flanked by the sun and moon, indicating that it stands at the centre of the world, where the moon does not wane and the sun does not set. Detail from a Persian bowl of the late Sassanian era (AD 226-641).*

*A depiction of shamanistic transformation into a bird-like figure by contemporary Cape Dorset Eskimo (1970).*

the ancient and archaic peoples of the earth: 'It is a universally disseminated idea connected with the belief in the possibility of direct communication with the sky.'[9] By the time they were examined by Western investigators, the nature of the shaman's flight among some peoples had certainly become obscured. The shaman might only be able to approximate the flight through ecstasy induced by narcotics of one sort or another, and there is a question, as it were, as to just how 'high' he got. But even where the shamans were degenerate and the magical flight may not have been attained in its truest and purest form, such individuals were at the end of an extremely ancient tradition which went back to the very earliest times when such flights were entirely real and the most important single experience in the life of the people.

## The Mysteries of Cosmic Flight in Early Religions and Cults

Thus the flight of Unas connects with the key element in archaic religious patterns around the globe. But that is far from all. It also illuminates the central experience of the classical Mystery Religions. These religions existed around the Mediterranean basin and to the east for thousands of years before Christ and for a few centuries afterwards. Despite variations in the names of the gods and goddesses, the cults all approximated a single type. The Egyptian rites of Isis and Osiris, the Persian rites of Mithras, the Greek Dionysian (or Bacchic) rites, the Orphic

Mysteries (a reformed Dionysian cult), and the Eleusinian Mysteries centred around Demeter and Proserpine (or Persephone) were all basically the same rites with the leading deities renamed and adapted to particular nations and times. Plutarch, in *Of Isis and Osiris* for example, points out that Demeter is the same as Isis and that Osiris 'is one and the same with Bacchus'.[10] It is widely recognized that Isis and Osiris also correspond to Ishtar and Tammuz of Babylonia.

The most generally acknowledged aspect of these cults is that they commemorate the death and resurrection of the god, the latter process facilitated by the ministrations of the goddess. Isis, for example, collects and reconstitutes the pieces of Osiris' body after it had been dismembered and the parts scattered by Set. Similarly, to retrieve the spirit of Tammuz at his death, Ishtar passed through the seven precincts of Hades, at each of which she was required to shed some of her clothing so that when she arrived in the presence of the queen of the infernal regions she was naked. Since these deaths and resurrections typically take place in the spring, the traditional scholarly interpretation has been that the rites centred around such divinities reflect the cycles of vegetation and fertility.

However, many of the great minds of classical antiquity such as Pythagoras, Plato, Cicero, Plutarch, Herodotus and others were initiates of these mysteries. This in itself indicates that these mysteries were something more than mere fertility cults, and recent scholarly interpretations have at least been alert to that possibility. In *Greek Folk Religion* Martin Nilsson notes:

It may be permitted to ask whether deeper ideas of life and death were not evoked by the Eleusinian Mysteries. Perhaps they were. In a remarkable fragment Pindar [the Greek poet] says: "Happy is he who, having seen this, goes beneath the earth; he knows the end of life and he knows its god-sent beginning."[11]

In *Patterns in Comparative Religion* Eliade went much further. He concludes that the myths did not arise from the patterns of nature but that:

The drama of the death and resurrection of vegetation is revealed by the myth of Tammuz, rather than the other way about. Indeed, the myth of Tammuz, and the myths of gods like him, disclose aspects of the nature of the cosmos which extend far beyond the sphere of plant life; it discloses on the one hand, the fundamental *unity* of life and death, and on the other, the hopes man draws, with good reason, from that fundamental unity, for his own life after death.[12]

The most famous site of these mysteries in the Greek world was at Eleusis about 12 miles northwest of Athens, where rites were held over a period of 2000 years. The rites were aptly termed 'mysteries' because anyone initiated into them was sworn to secrecy not to reveal their contents. The dramatists Aeschylus and Aristophanes were nearly slain for having revealed too much in their plays. Under these circumstances, the central experience of the Eleusinian Mysteries was a successfully kept secret for thousands of years.

In spite of the oath of secrecy, in the second century after Christ a zealous devotee of Isis provided a glimpse of his initiation, which he attained after baptism, a period of fasting and much expense in payment to the priests. In *The Golden Ass* Lucius Apuleius discloses just enough of the culmination of his experience for us to see that it is another account of a mystical flight:

> Then the High Priest ordered all uninitiated persons to depart, invested me in a new linen garment and led me by the hand into the inner recesses of the sanctuary itself. . . . If I were allowed to tell you . . . you would soon hear everything; but . . . my tongue would suffer for its indiscretion. . . . I will record as much as I may lawfully record for the uninitiated. . . . I approached the very gates of death and set one foot on Proserpine's threshold, yet was permitted to return, rapt through all the elements. At midnight I saw the sun shining as if it were noon; I entered the presence of the gods of the underworld and the gods of the upper-world, stood near and worshipped them.[13]

Lucius then remarked: 'Now you have heard what happened, but I fear you are still none the wiser.' If we had his account alone, we wouldn't be. But we recall that Unas was described as if he was orbiting the earth; he was told to follow the sun and he went 'around the sky like Re'. Black Elk was taken to 'the place where the sun continually shines'. 'At midnight' Lucius 'saw the sun shining as if it were noon'. It very much seems that the core experience of initiation into the classical mysteries was in broad terms the same experience of Osirian king-initiates thousands of years earlier, and of native American shamans as late as the early 20th century.

If we look further, we soon find that these flights are features of most of the great religions. Of course in the Old Testament there is Jacob's Ladder (GENESIS 28) with the angels of God ascending and descending on it. In the Second Letter to the Corinthians Paul reports: 'I know a man in Christ who . . . was caught up to the third heaven—whether in the body or out of the body, I do not know.' Many have speculated that it was Paul himself who had this experience. Further, both Buddha and Muhammad also ascended to heaven. Buddha took 'seven steps' to the north 'in a single direction', and declared 'I am at the top of the world'.[14] Muhammad followed the same path. In a recent study J. R. Porter has remarked: 'In the course of his ascension, Muhammad passes through seven heavens and the notion of heaven as divided into seven levels which the shaman visits on the course of his initiatory journey is again a very common one.'[15]

*A medieval Persian representation of the order of the spheres, seven in number. Note the ladder leading to the first heaven and the trees in each sphere placed on the central axis.*

## The Phenomenon of Flights Which Transcend Time and Culture

The mystical flights recorded in various scriptures and traditions follow the same pattern as the core event in ancient initiations. It becomes clear that we are dealing here with a phenomenon of fundamental importance that transcends culture and time. It is always the same experience of a flight to heaven or the underworld—an underworld which is not in the bowels of the earth but outside and beyond the earth, a region in space opposite heaven. Examining classical writers, it becomes apparent that part of the reason for this remarkable uniformity is that initiation is linked to another phenomenon transcending culture and time—it is, simply, that we all die.

'To die is to be initiated', Plato wrote,[16] and in his essay, *'On the Soul'*, Themistios makes it a little clearer: 'The soul at the point of death has the same experience as those who are being initiated in great mysteries.'[17] As Lucius Apuleius put it: 'The rites of initiation approximate to a voluntary death from which there is only a precarious hope of resurrection.' Those who succeeded are, he says, 'in a sense, born again and restored to new and healthy life'.[18] In other words, Apuleius merely followed the steps of Ishtar and similar figures who triumphed over death by venturing into the after-death realms and then returned to the land of the living. This same pattern of voluntary death and resurrection applies to shamanism and initiatory rites of archaic peoples everywhere. In those cases where the flight occurred spontaneously, the shaman lay as dead during the crisis; he could not heal others until he was resurrected and had healed himself.

In the African secret society of the Bakhimba, the initiation ordeals continued for from two to five years. Among these people, according to Eliade:

> The neophyte has to be "killed", a scene which is acted at night, and the old initiates "sing . . . the dirge of the mothers and relatives over those who are about to die". The candidate is scourged, and drinks for the first time a narcotic beverage called the "drink of death", but he also eats some calabash seeds, which symbolize the intelligence—a significant detail, for it indicates that through death one attains to wisdom. When he has drunk the "drink of death" the candidate is seized by the hand, and one of the old men makes him turn round and round until he falls to the ground. Then they cry, "O, so-and-so is dead!" (and) intone a funeral dirge: "He is quite dead, ah! he is quite dead! . . . I shall never see him again!"
>
> And in the same way, in the village, he is also mourned by his mother, his brother and his sister. Afterwards "the dead" are carried away, on their backs, by relatives who are already initiated, into a consecrated enclosure called "the court of resurrection". There they are laid, naked, in a grave dug in the form of a cross, and remain there until the dawn of the day of . . . "resurrection".[19]

*A scene from the coffin of Hent-Taui (21st dynasty, 1113-949 BC) interpreted by N. Rambova as 'resurrection as the birth child in the solar barge, shows not only a connection between death in one realm and rebirth in another, but also between travel in the upper realm leading to restoration and resurrection in the world below'.*

*Detail from a Mayan fresco at Bonampak, Mexico showing the widespread association of high priests and initiates with feathered and flying creatures. Here a priest is being arrayed as the plumed serpent Kukulcan, or Quetzalcoatl.*

*During the peyote ritual, incense is scattered over the flame.*

The pattern becomes still clearer and more provocative in the light of contemporary research. We can see today that the purpose of most initiatory rituals was to induce the very physical crisis which almost invariably accompanies spontaneous out-of-body flights. We have suggested that the sarcophagi in the Egyptian pyramids were used to hold the bodies of initiates during these rituals. It is interesting that in Mexican legend the great Quetzalcoatl was also interred in a stone sarcophagus. He was said to be so confined for four days in order to reestablish his celestial origins.[20] Among various American Indian peoples preparatory rites included lengthy seclusions in smoke huts and sweat huts. Before and during these confinements, the candidates 'purified' themselves, principally through fasting and/or through rites like baptism. Some peoples used psychotropic substances — usually in conjunction with fasting and confinement — to induce altered states of consciousness which made the flight more accessible. In Crete a terracotta goddess was found with three large opium bulbs sprouting from her head, and in Lebanon a great jar of marijuana was uncovered in a temple ruin.[21] Ancient Mexican peoples, the Maya in particular, made use of peyote, hallucinogenic mushrooms and other plants as aids in communicating with the gods.[22] Such plants were considered sacred and god-given.

There is substantial, contemporary evidence for the efficacy of these initiatory techniques. Experiments into sensory deprivation and altered states of consciousness have been conducted by governments and universities since the 1950s. Interest in sensory deprivation seems to have been sparked by the brain-washing problem encountered during the Korean War, but was later expanded because of its relevance to the psychophysiological aspects of space flight.

A typical instrument in some sensory deprivation research is a specially constructed box the size of a large coffin, in which the subject is isolated from external stimuli. In other experiments the isolation chamber is the size of a small room, sometimes partially filled with water to minimize the effect of gravity. The subject may wear a special suit and other gear to further cut him off from outside influences.

Probably the most popularly known researcher in this area has been John C. Lilly, a medical doctor and psychoanalyst who in addition to years of study on solitude and confinement has also worked extensively in biophysics, neurophysiology, electronics and neuroanatomy. Lilly and others have found that such isolation alone, if endured for more than a few hours, often results in altered states of consciousness and hallucinations. The subject is also far more susceptible to hypnotic suggestion. Both conditions are precursors of out-of-body experiences.

*The granite sarcophagus in the King's Chamber of the Great Pyramid invites comparison with a contemporary sensory deprivation box.*

One has only to note the striking similarity between some of these isolation chambers and the sarcophagi in Egyptian pyramids to see that the sarcophagi, too, could have served the same purpose. Moreover, with their lids in place and an initiatory ritual that may have taken several days, the sarcophagi also limited the amount of oxygen available to the initiate. As Stanislav Grof and Joan Halifax remark:

It is well known from many different sources that a limited supply of oxygen or an excess of carbon dioxide produces abnormal mental states. Experiments with the anoxic chamber have shown that lack of oxygen can induce unusual experiences quite similar to LSD.[23]

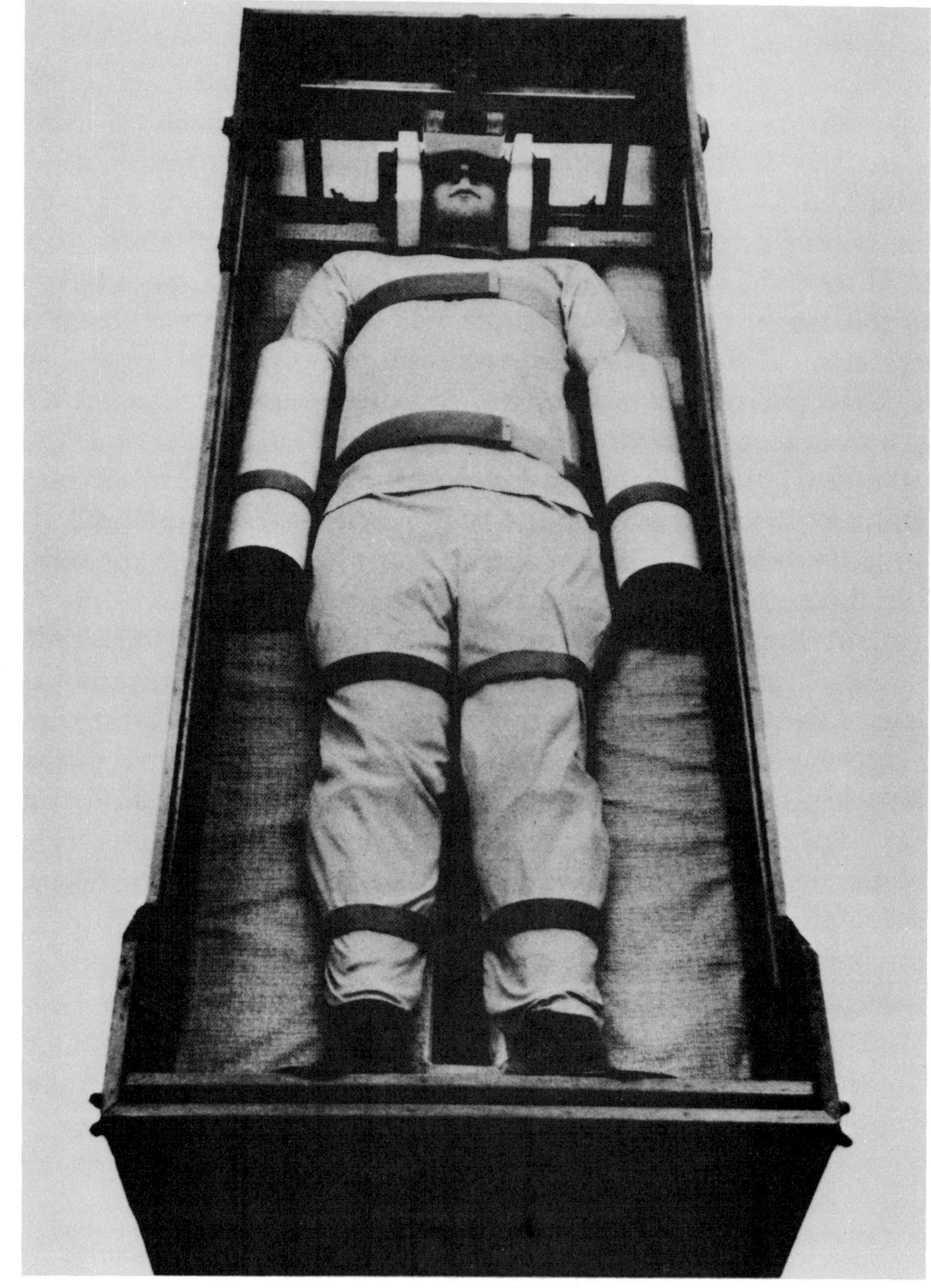

*A sensory deprivation box developed at the University of Manitoba.*

Lilly, who had had several spontaneous out-of-body flights previously, found that LSD could be used as a launching pad for such travels, but concluded that it was better to achieve them organically. Eliade reports that the use of drugs to induce trances was a degenerate stage in shamanism.[24] Nevertheless, it can easily be seen how prolonged seclusion in a sarcophagus, smoke hut or sweat hut such as the Indians used, all made use of the anoxic effect. The connection between contemporary LSD research and the caches of opium and marijuana found at ancient sacred sites requires no further comment.

In the light of recent research, again we find that many archaic practices that once seemed benighted and barbaric are actually based on sound physiological and psychological principles. It is now possible to see that the mystical flights of these peoples were originally genuine out-of-body experiences and that the cultures which pivoted around their attainment had explored dimensions of man and the universe we in the West are only now rediscovering.

There is a single great pattern that emerges in ancient and archaic initiations, in the careers of the ancient gods, and in the lives of the great religious leaders: a voluntary death, a flight from the body, descent into the underworld and/or ascent into heaven; then a return to and resurrection of the body. Not only is the pattern the same, the symbols within the pattern are also amazingly consistent. The celestial water described by Black Elk and Unas is widely reported; otherwise it is something like 'the balm of life' in Norse mythology. The mother of Lemminkäinen, a slain god, sends a bee beyond the ninth heaven to obtain and bring to her the balm of life that she may resurrect her son.[25]

Like Osiris and Dionysus, Lemminkäinen was dismembered. The very general pattern in the initiatory visions of shamans and their like was to see themselves dismembered, or at the least in some way mutilated. In the legends surrounding the ascent of Muhammad, he was said to have seen his chest opened, his heart removed and a drop of black blood wrung from it. The initiate experienced the death of the body on two planes: by leaving the physical body in a death-like swoon and by witnessing a vision of its destruction or mutilation. Following this double surrender, he receives his body again, reconstituted and rejuvenated. Like the division of the universe into three cosmic zones of sky, earth and underworld united by an axis of some kind, the connection of out-of-body flight with death and rebirth is amazingly universal.

Except when the transforming vision occurred spontaneously, initiations were usually staged in secluded places, in the dim inner sanctuaries

*The sun dance performed at the summer solstice may also be performed when the community feels a need for renewal of strength and vision. According to Black Elk, the dancer prepared himself with prayer, fasting and purification in a sweat lodge before submitting himself to have strips of rawhide from the tree thrust through cuts in his flesh. He would then dance as long as he could stand the pain, or until the flesh tore away, a ritual promoting visions and out-of-body flight.*

*In this contemporary depiction of the sun dance, the artist has incorporated the symbolism of the cross and cosmic axis with the sacred tree, which is aligned to the sun setting behind a mountain. Note the two small figures attached to the top of the tree.*

*The Ploutoniaon, a sacred cave at Eleusis in southern Greece, which was a centre of mystery religions for over 2000 years.*

of temples, in caves and grottoes in the inner recesses of the earth and, where these were not available, in chambers under artificial mounds. The Egyptians themselves referred to the pyramids as 'mountains' and of course they enclose artificial 'caverns'. What these were to the Nile people, New Grange (an artificial mound) and St Patrick's Purgatory (a natural cavern) were to the Celts; as were the *kivas* (with a hole to the underworld) to the Hopi and Cliff Dwellers of the American Southwest. Sites with parallel functions are found worldwide.

## The Conquest of Death

Initiations and the accompanying mystical flights accomplished three primary purposes. The first and basic purpose of the ancient mysteries was to impress upon the initiate the fact that he consisted of something else

*The reconstructed interior of a small mound at Fourknocks, County Meath, Ireland. Note the decoration of the stone above the small recessed area.*

*A small artificial mound known as Bryn Celli Ddu, in Anglesey, northwest Wales.*

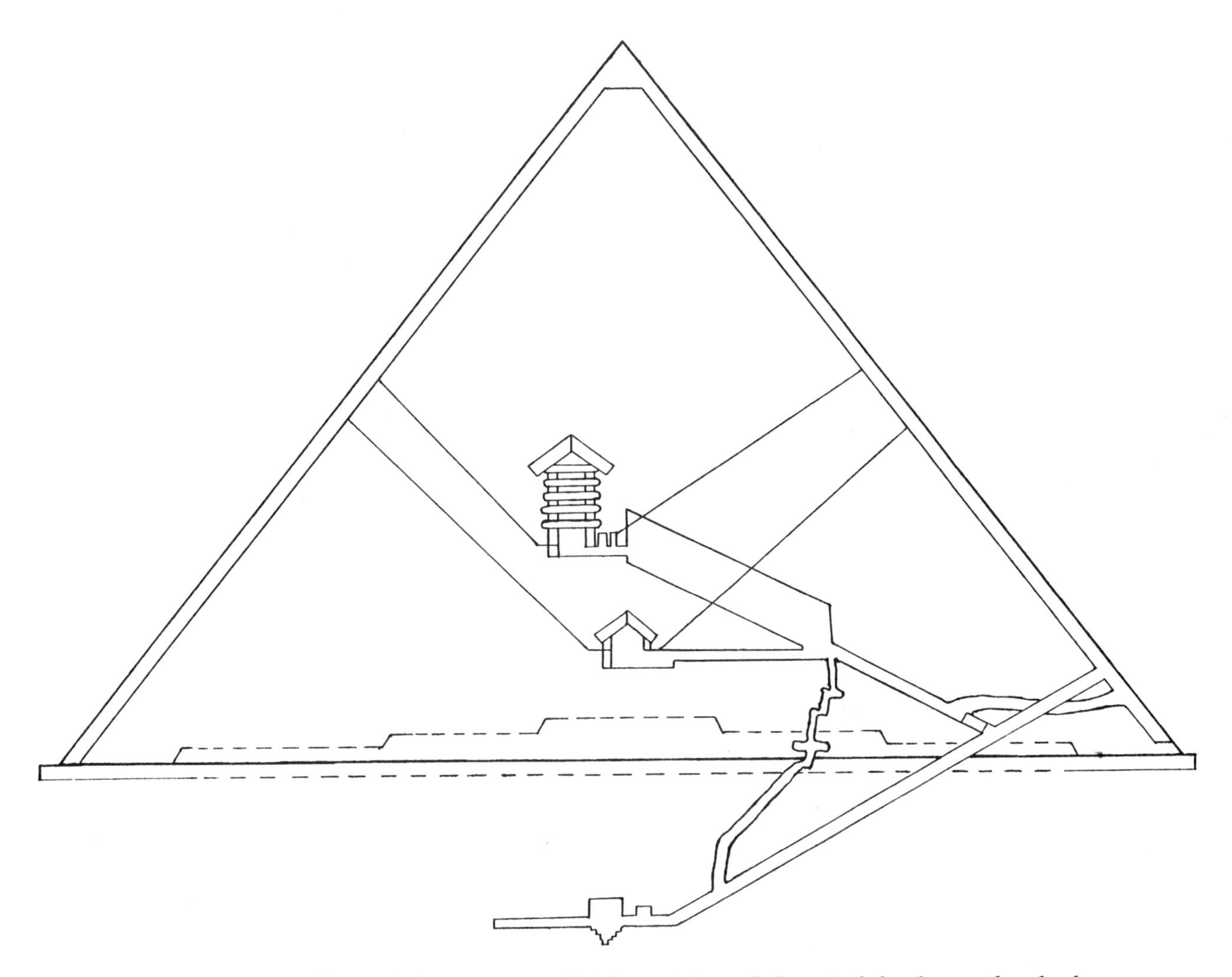

*The Great Pyramid may also be seen as an artificial mountain, enclosing sacred chambers on three levels.*

than merely his physical body. He rehearsed his own death in leaving the body, and experienced continuing consciousness separated from it. In returning to his body which had been left in a death-like state and re-enlivening it, he simultaneously knew a form of resurrection. One of the invariable benefits procured by successful initiates was that they no longer feared death.

Although long shunned by some scientists, the examination of death and what may survive it has lately advanced dramatically through the work of individuals like Elizabeth Kübler-Ross, Karlis Osis and Raymond Moody. Working with the dying and hearing first hand the spiritual adventures of those who had nearly died, Kübler-Ross has declared: 'I already have more than enough evidence that there is an afterlife.'[26] Patients who wandered past the point of medical death, had some form of otherworldly experience and then, like Jung, returned to their bodies, not only no longer feared death but looked on life with new eyes. As it was expressed by a 17th century Augustinian monk named Abraham a Sancta Clara: 'The man who dies before he dies, does not die when he dies.'

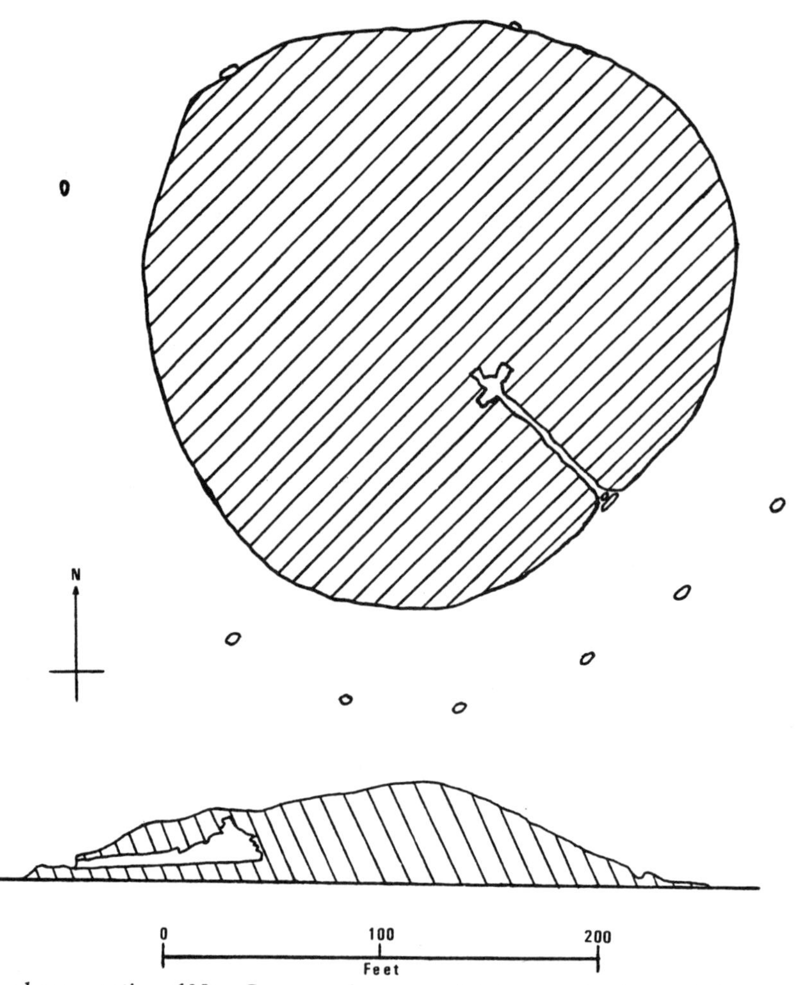


*Ground plan and cross section of New Grange, a famous artificial mound in the bend of the Boyne River in Northeastern Ireland. It is here, legend tells us, that the sons of a king of Ireland came to fast for three days and nights on a quest to communicate with the Underworld. Like many other sites associated with initiatory ritual, New Grange also has astronomic significance, the passageway being so orientated that sunlight strikes into the heart of the chamber for 20 minutes only on the shortest day of the year (winter solstice). New Grange is thought to date to the third millennium BC.*

*The discovery of human bone fragments in the mound indicates that the sacred site may have been appropriated for burial purposes in later times. The presence of Roman coins and jewellery show that offerings may have been made at the site as recently as the fourth century AD.*

Commenting on the experience of the mysteries, Cicero wrote: 'We at last possess reasons why we should live, and we are not only eager to live, but we cherish a better hope in death.'[27] And an Eleusinian initiate declared: 'It is a fair mystery that comes to us from the Blessed; for mortals, death is no more an evil, but a bliss.'[28]

The second major purpose that could be achieved through initiation was the release of tremendous power: the power 'to make live and to destroy', the power to control the weather (rainmaking) and the behaviour of animals, the power to heal and to see the future. The initiate returned from his journey a completely altered man; often his name was changed; and if he had ascended high enough he became a channel for the powers of life — as had Black Elk.

*In the three side chambers of New Grange are large bowls or basin stones. This granite basin stone, 1.15 metres in diameter, rests upon a larger stone in the eastern chamber. Such stones, found at various megalithic sites throughout Ireland, are similar to the stone troughs or stone beds of later Christian saints and hermits of the area, probably corresponding in function to the sarcophagi of the Egyptian pyramids. The initiate may have sat cross-legged in the basin.*

*The entrance to New Grange. The giant spiral-marked stone, doorway and astronomic window above the door are all original. The facing material is a recent reconstruction.*

*Triple spiral cut in the wall of the main chamber of New Grange. The spiral leading to an opening at the centre is an ancient symbol for the journey to the other world. Among the most common descriptions of contemporary out-of-body travel is a passage through a spiral or tunnel with a light at the end.*

*The passage leading to the inner chambers of New Grange.*

*The Cliff Dwellers' colony at Mesa Verde, Colorado. Each colony had several men's societies, each of which had its own* kiva *or sacred underground room.*

*A* kiva *at Mesa Verde. This circular room served primarily as an initiation chamber. Note the altar and the small hole, lower left, a symbolic connection to the Underworld and the original home of man.*

*Voyage of the Toltec Initiate. A sophisticated map of consciousness depicting the cycle of birth, death and rebirth in a form simultaneously suggestive of a seven-chambered cavern and a cosmic tree. Below right is an embryonic figure in a womb-like enclosure, signifying Becoming, which is baptized from above with fire and water. The two lower figures are gods, the vortices between them their words. The one who wishes to follow the path of the gods must pass through their words. The footprints of the gods run in both directions, showing that the path leads not only to seven different circuits of experience beyond earth, but out again to reenter this world from the centre, through which all must pass.*

*A North American Algonquin shaman as depicted by an Army officer around 1880.*

Many I cured with the power that came through me. Of course it was not I who cured. It was the power from the outer world, and the visions and ceremonies had only made me like a hole through which the power could come to the two-leggeds. If I thought that I was doing it myself, the hole would close up, and no power could come through.[29]

In Egypt thousands of years earlier, the Osirian king-initiates were similar channels of energy, perhaps far greater than Black Elk. Unas, we are told, made things green. Upon his return the text reads: 'Let the hoers be in jubilation.' [30] When the text declares: 'Unas is the lord of seed, he who takes the women from their husbands, wherever Unas wants, according to the wish of his heart', it is not a description of a haughty barbarian, but of the duties and prerogatives of one of the semi-divine kings of deep antiquity.

He was not to be envied. The king was the centre not only of the government but of the religious system as well. As God stands to the world, so the king was to his kingdom. This macrocosmic-microcosmic pattern was the source of the concept of the divine right of kings, and of the king's enormous responsibilities. He was understood to be the nation's link with higher realms. The function of his initiation was to reaffirm that connection. Such a journey as he undertook was not necessarily a matter of choice. During later times, the tradition of the Heb Sed Festival required that at least once every 30 years he don the robes of Osiris and enact the death and rebirth of the god in order to ensure the renewal of both himself and the nation.

Some pharaohs seem to have relished these renewals. After his first celebration of the Heb Sed on his 30th anniversary, Ramses II performed the festival eight more times in the next 23 years.[31] On the other hand, if the king were particularly inept or disliked by enough of the priests, it is not difficult to imagine that he might find it was made impossible for him to arise from the sarcophagus. Even without political complications, the journey upon which he was sent might not be completed successfully by everyone. If improperly trained or otherwise unprepared, the initiate might return demented, or he might not return at all.

Belief in the intimate relationship between all things in nature and the affairs of men made the king ultimately responsible for everything from the country's military security to the weather and the state of the crops. This interrelationship was the basis for the importance that was attached to the study of omens in many ancient nations. In China a collection of ancient manuscripts known as the Great Law was compiled about 1050 BC in which this relationship is clearly explained:

*A comet.*

*A Renaissance depiction of the otherworldly vision associated with the breakthrough in plane.*

> It is the duty of the government all the time to watch carefully the phenomena of nature, which reflect in the world of nature the order and disorder in the world of govenment. The government is bound to watch the phenomena of nature in order to be able at once to change what is in need of change. When the course of nature runs properly, it is a sign that the government is good, but when there is some disturbance in nature it is a sign that there is something wrong in the government. . . . Any disturbance in the sun accuses the emperor. A disturbance around the sun accuses the court and the ministers. A disturbance in the moon accuses the queen and the harem. Good weather that lasts too long shows that the emperor is too inactive. Days which continue to be cloudy show that the emperor lacks understanding. Too much rainfall shows that he is unjust; lack of rain shows that he is careless. . . . A good harvest proves that all is well; a bad harvest that the government is at fault.[32]

In Egypt the same principles applied. If the king were to be the channel of blessings to the nation from on high, like Black Elk he could only transmit those energies if he had made contact with them. Thus the third and highest state of initiation was the attainment of cosmic consciousness in which one meets the divine within and without. 'Thou hast become being,

*Unusual movements in the heavens inspired special interest among those who sought to divine the future from signs in nature.*

thou art become high, thou art become spirit! Cool it is for thee in the embrace of thy father, in the embrace of Atum.' 'And while I stood there I saw more than I can tell and I understood more than I saw; for I was seeing in a sacred manner the shape of all shapes as they must live together like one being.' It was with good reason that the kings of old pursued the paths of heaven. Their knowledge, experience and powers were real. Their authority was, for many centuries, unchallenged.

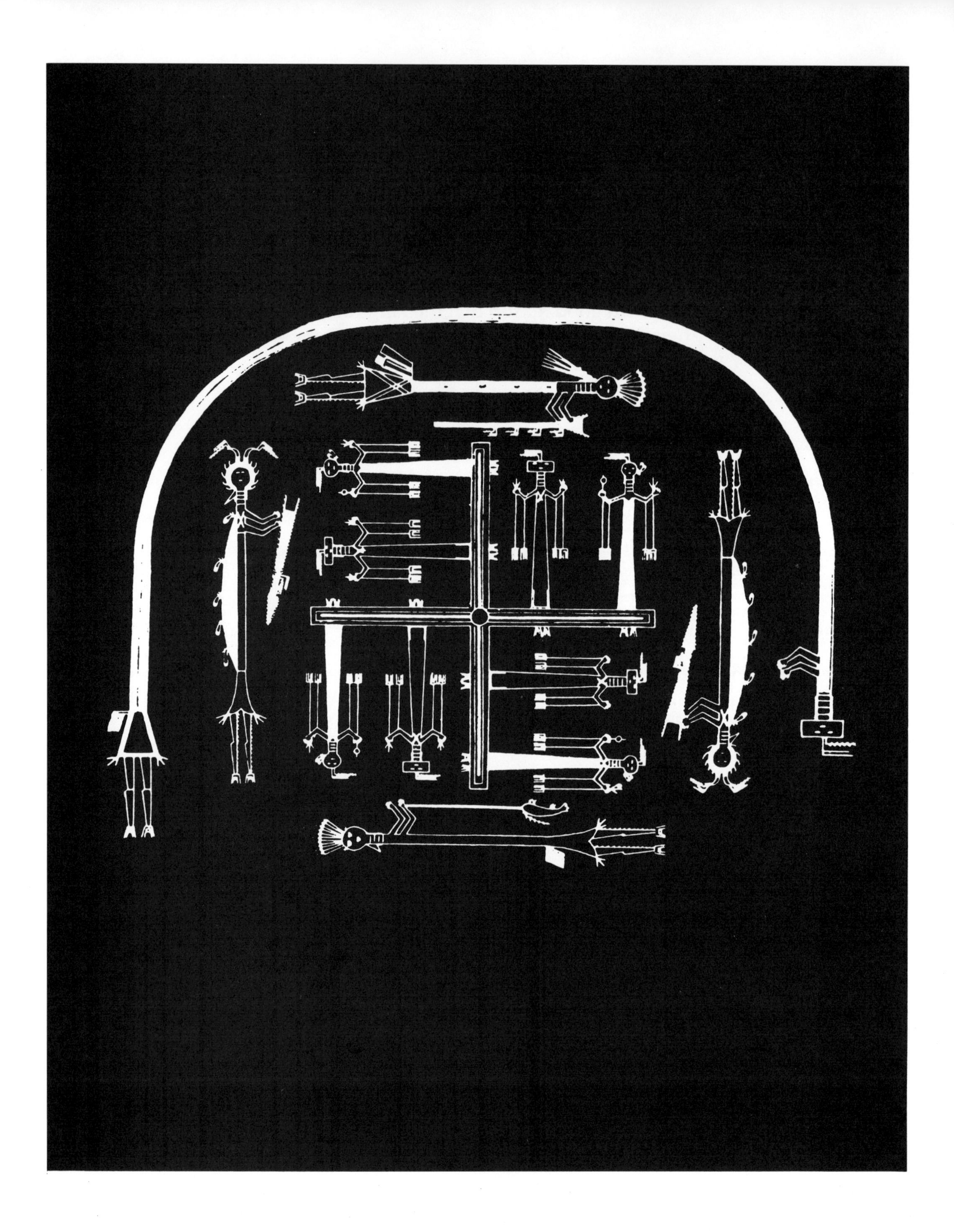

CHAPTER EIGHT

★————————————————★

# THE STAR SHIPS

## The Case for the 'Desire Body'

In unravelling all the implications of the flight of Unas and in searching out why he went to the circumpolar stars, we must necessarily heed the first lesson of the ancient mysteries: man consists of something else than merely his physical body.

While in the modern world this 'something else' has long been seen as especially elusive and doubtful, the ancient world produced numerous and varied descriptions of it. According to St Paul in I Corinthians, 'there are celestial bodies and there are terrestrial bodies'. Among the simplest and most poetic conceptions are the 'eternal bodies' defined in a fragment of the literature attributed to Thrice Greatest Hermes:

> The Lord and Demiurge of all eternal bodies . . . when He had made them once for all, made them no more, nor doth He make them now. Committing them unto themselves, and co-uniting them with one another, He let them go, in want of naught, as everlasting things. If they have want of any, it will be want of one another and not of any increase to their number from without, in that they are immortal.[1]

Here the conception seems to be that man's eternal body is a simple single soul or spirit. In other ancient sources, however, the something else in man appears to be neither simple nor single. Sifting the terminology of different cultures, sects and times, it seems that most ancient and archaic peoples believed that man consists of multiple invisible elements and components.

*The sky goddess Nut lifted from the earth god Geb, as seen in the 21st dynasty (1113-949 BC) Papyrus of Nisti-ta Nebet-Taui. Here the role of Shu, god of the air, is taken by an ape-headed god, possibly in the form of the Bentui who traditionally greets the appearance of Ra.*

In addition to the physical body and the kind of immortal soul described by Hermes, many peoples describe what might be called a magnetic duplicate of the individual that possibly survives the death of the physical body for a while, yet never leaves the earth and eventually dissolves. The Egyptians seemed to have called this the *ka*, [2] the aboriginal Hawaiians the *kino aka*, or in the terms of the chronicler of Hawaiian shamanism, M.F. Long, 'a low shadowy body'.[3] In German the word *doppelgänger* sometimes apparently describes the same element.

Typically, there are usually at least two higher bodies above this low shadowy body. The psychology of Hawaiian shamanism defines man as having three levels of consciousness — the unconscious or subconscious, the normal waking consciousness and the superconscious. The low shadowy body is the vehicle of the subconscious. The shadowy body of the conscious self is interblended with the corresponding element of the subconscious, and though a separate element, is also called *kino aka*. This may be what the Egyptians called the *ba*. The superconscious, the highest element in man, is said to be definitely immortal and is referred to by Long as a 'dual-sexed parental spirit', a kind of ultimate personal guardian and repository of consciousness that he found similar to the guardian angel of Christianity and Islam. It resides in its own very light body. The Egyptians also defined higher bodies beyond the *ka* and *ba*, called the *sa* and the *khu*, which seem to partially match the Hawaiian conception.

Of course, our own ignorance and the lack of uniform terminology on the part of ancient writers and their commentators alike make it impossible to arrive at a single complete system. However, virtually all archaic and ancient peoples believed in some kind of soul or spirit and often defined it as multiple in structure. It is intriguing that even with this most elusive of subjects, recent research into this long neglected area has again produced results directly comparable to classical descriptions. In *The Tibetan Book of the Dead* the deceased is instructed that even if when living he had been 'blind of the eye, or deaf, or lame, yet on this After-Death Plane thine eyes will see forms; and thine ears will hear sounds, and all other sense-organs of thine will be unimpaired and very keen and complete.'[4] He is further told that in this after-death plane he travels in a 'desire body', a body commanded by one's wishes. This 'is not a body of gross matter, so that now thou hast the power to go right through any rock-masses, hills, boulders, earth, houses and Mt Meru itself without being impeded.'[5] The translator, W.Y. Evans-Wentz, points out that this power, supernormal in the human world, is normal in the 'fourth dimensional after-death state.'[6]

*In the Papyrus of Nefer-ubenf (about 1000 BC) a man is depicted as consisting of two post-mortem elements: a soul in the form of a bird and a shade. They are outside a tomb, drawn to symbolize the horizon and the earth. It is surmounted by a sun disc.*

This description of travelling in a subtle body capable of unimpeded motion is exactly what we find in many of the accounts of those who passed the point of medical death, journeyed out of the body and returned. They are forever amazed that they can pass straight *through* walls and even other people, or that they have become invisible and normal living people pass *through* them. An accident victim reported:

People were walking up from all directions to get to the wreck. I could see them, and I was in the middle of a very narrow walkway. Anyway, as they came by they wouldn't seem to notice me. They would just keep walking with their eyes straight ahead. As they came real close, I would try to turn around, to get out of their way, but they would just walk *through* me.[7]

A woman described leaving her body as she was attended by a doctor and nurses while lying in a hospital:

I became separated. I could then see my body ... I stayed around for a while and watched the doctor and nurses working on my body, wondering what would happen ... I was at the head of the bed, looking at them and my body, and at one time one nurse reached up to the wall over the bed to get the oxygen mask that was there and as she did she reached *through* my neck ... [8]

Those that escape the body and have a specific goal they wish to reach usually fly at the speed of thought in a straight line and seem to be guided only by that mental goal. The aptness of the phrase 'desire body' is demonstrated in the account of an American medical doctor and psychiatrist, George Ritchie, who in *Return from Tomorrow* describes what he saw in December 1943 when as a 20-year-old he died of double-lobar pneumonia and nine minutes later returned to life. At first he did not know that he had died:

Almost without knowing it I found myself outside, racing swiftly along, travelling faster in fact than I'd ever moved in my life ... Looking down I was astonished to see not the ground, but the tops of mesquite bushes beneath me ... My mind kept telling me that what I was doing was impossible, and yet . . . it was happening. A town flashed by beneath me, caution lights blinking at the intersections. This was ridiculous! A human being couldn't fly without an airplane—anyhow I was travelling too low for a plane ... An extremely broad river was below me now . . . and on the farthest bank the largest city I had come to yet. I wished I could go down there and find someone who could give me directions. Almost immediately I noticed myself slowing down ... Finding myself somehow suspended fifty feet in the air was an even stranger feeling than the whirlwind flight had been . . . Down the sidewalk toward [an] all-night café a man came briskly walking ... I thought I could find out from him what town this was ... Even as the idea occurred to me—as though thought and motion had become the same thing—I found myself down on the sidewalk ... [9]

A subtle body that can pass through matter and fly with the speed of thought is indeed a vehicle in which man could cross cosmic distances.

## The Concept of Self-Judgment After Death

In truth, modern man can learn much from the Tibetans, ancient Egyptians, Greeks and many others about these other dimensions of man and the universe. A typical modern reaction to leaving the body was voiced by a young woman:

I thought I was dead, and I wasn't sorry that I was dead, but I just couldn't figure out where I was supposed to go. My thought and my consciousness were just like they are in life, but I just couldn't figure all this out. I kept thinking "Where am I going to go? What am I going to do?" and "My God, I'm dead! I can't believe it!"[10]

It was precisely because people remain conscious after death and for the purpose of answering such questions that ancient and archaic peoples all over the earth recorded cosmologies which are 'maps of consciousness' for both the living and the dead. In *The Human Encounter with Death*, psychiatrists Grof and Halifax remark:

We are now beginning to learn that Western science might have been a little premature in making its condemning and condescending judgments about ancient systems of thought. Reports describing subjective experiences of clinical death, if studied carefully and with an open mind, contain ample evidence that various eschatological mythologies represent actual maps of unusual states of consciousness experienced by dying individuals. Psychedelic research conducted in the last two decades has resulted in important phenomenological and neurophysiological data indicating that experiences involving complex mythological, religious and mystical sequences before, during and after death might well represent clinical reality.[11]

Today, of course, there seems to be an incredible variety of religions, creeds and sects, the adherents of which sometimes believe that theirs is the only way. Yet beyond the multiplicity of names and forms there are patterns, powers, figures and roles in religions and cosmologies from all ages that are remarkably similar. Part of the clinical reality behind virtually all creeds is that upon death of the physical body, the surviving element in man faces judgment. The image that the soul or heart is weighed in a balance cuts across cultures and millennia. The earliest Egyptian conception of the judgment is almost identical to that sometimes expressed by the classical Greeks and to judgment scenes of medieval Christianity. The scales in which the heart or soul is weighed correspond to the mirror in which the Buddhist expects to see the record of his life. *The Tibetan Book of the Dead* explains that the actual judgment is an act performed by some element in the individual himself:

> You are now before Yama, King of the Dead. In vain will you try to lie, and to deny or conceal the evil deeds you have done. The judge holds up before you the shining mirror of Karma, wherein all your deeds are reflected. But again you have to deal with dream images, which you yourself have made, and which you project outside, without recognizing them as your own work. The mirror in which Yama seems to read your past is your own memory, and also his judgment is your own. It is you yourself who pronounce your own judgment, which in its turn determines your next birth.[12]

This is indeed the clinical reality reported by Moody and others: the judgment is not made by some other being, but 'within the individual'.[13] What is most remarkable is that the clinical reality of the 'actual maps of consciousness' as Grof and Halifax call them, is not merely a 'mystical sequence' in some vague psychological system encountered by the soul after judgment. The extraordinary thing is that these maps of consciousness point to a great underlying pattern which is partly expressed in the physical, visible universe—in the outer world of the planets and the stars.

*In this Egyptian depiction of the weighing of the heart by Anubis in the Papyrus of Anhai, the symbol of the heart is shown hanging from the Feather of Truth and balanced against the goddess of truth in the scales. The baboon surmounting the scales represents Thoth; the grisly creature on the left — part crocodile, part lion and part hippopotamus — is the Devourer of Damned Souls.*

*Judgment scene from a Greek vase: Hermes weighs the fate of two heroes as Zeus looks on. Although the Greeks are not usually credited with a belief in the post mortem weighing of the soul, miniature scales with butterflies, symbols of the soul, have been found in Mycenean tombs.*

*Early 12th century mural in the Church of St Peter and St Paul, Chaldon, Surrey. At the upper left, St Michael weighs souls with scales just as the Egyptians depicted judgment thousands of years earlier. Souls ascend the eight-runged ladder of salvation. At the upper right, Christ redeems Old Testament souls from hell. Below are the torments of the damned.*

CHAPTER NINE

# THE RIVERS OF HEAVEN

## Where Does One Go After Death?

To die may be to be initiated, but death itself does not automatically produce wisdom, or even a clear view of the voyage ahead. Indeed, all the indications are that when in the desire body even one's illusions may take on life to amaze and dismay the wanderer.

Perhaps one of the greatest hazards the spirit faces once it has finally separated from the physical is the inability to understand its altered state. Contemporary reports indicate that there are a lot of bewildered people on 'the other side'. If the individual was deeply attached to affairs and concerns of the earth at the moment of death, he literally may not know that he has died, or even if aware of that much, finds being out of the body so totally unexpected that he may long focus downward on affairs on earth instead of upward, and thus remain in earthbound fascination.

The extraordinary thing is that many navigators before us have made maps of where we travel, and the realms that appear on some of these charts form a pattern whose incredible implications seem not to have been noticed. A middle-aged man who had had a cardiac arrest related:

> I had heart failure and clinically died. . . . I remember everything perfectly vividly. . . . Suddenly I felt numb. Sounds began sounding a little distant. . . . All this time I was perfectly conscious of everything that was going on. I heard the heart monitor go off. I saw the nurse come into the room and dial the telephone, and the doctors, nurses, and attendants came in.

*After judgment upon death, one of the Osirian dead travels in a barque on a canal or river in a celestial landscape including the moon, sun and apparently a third heavenly body, as shown in a late* Book of the Dead *papyrus, about 1000 BC.*

As things began to fade there was a sound I can't describe; it was like the beat of a snare drum, very rapid, a rushing sound, like a stream rushing through a gorge. And I rose up and I was a few feet up looking down on my body. There I was, with people working on me. I had no fear. No pain. Just peace. After just probably a second or two, I seemed to turn over and go up. It was dark — you could call it a hole or a tunnel—and there was this bright light. It got brighter and brighter. And I seemed to go *through* it.[1]

During another medical crisis, a woman saw the same tunnel and light: 'And after I floated up, I went through this dark tunnel ... I went into the black tunnel and came out into brilliant light . . . '[2] Others called back from the other side have seen a river. An elderly victim of a heart attack recounted: 'You just can't imagine it. When you get on the other side, there's a river. Just like in The Bible, "There is a river ..." It had a smooth surface, just like glass . . . '[3] And in another case:'The next thing I knew it seemed as if I were on a ship or vessel sailing to the other side of a large body of water.'[4]

A great many of Moody's reports refer to buzzing, ringing or drumming sounds at the moment of death. Passages such as tunnels and holes, the light at the other end, and then boundaries such as rivers appear again and again in contemporary descriptions of transition into the other world. It is interesting and somewhat extraordinary that these same elements also appear in many other sources spanning thousands of years and many cultures. The tunnel or hole corresponds to the hole through which the soul of the shaman flies, often with the aid of drums and bells, and also reminds us again of the Bible's Valley of the Shadow of Death. In *The Tibetan Book of the Dead* the dying are given repeated instructions not to fear the sounds of drums, timbrels and trumpets but to follow the clear light.

Of course, not all contemporary accounts follow the same pattern. Some of these scouts do not get past simply floating in a peaceful darkness. Others swoon or lose consciousness at the moment of death — a common occurrence to be avoided if possible according to the Tibetans—and simply 'wake up' in other realms where they report an all-encompassing 'golden light', paradisiacal landscapes and other beings, and 'cities of light' or borders and thresholds which, they are told, they are not to enter if they wish to return to the earth.

Trying to locate these scenes on a larger map of consciousness is all but impossible working from strictly contemporary sources—as impossible as trying to map China from the scenes reported by Marco Polo. However, in the modern accounts collected by Moody, Grof, Osis and others there are many parallels with the realms beyond described by the Egyptians, Hindus, Buddhists, Tibetans, Plato and Dante. The rivers of heaven are mentioned in these and numerous other archaic and classical descriptions. Charting these rivers as they appear in these sources, a definite and remarkable structure of the other world emerges. This structure is of sufficient scope to accommodate the variety of modern visions of what lies beyond the tunnel—and identification of this structure leads to some quite extraordinary implications.

The most familiar map of the other world in Western civilization is that of Dante as he described hell, purgatory and heaven in the *Divine Comedy*. Heaven, hell and purgatory are each divided into nine realms and these are typically divided by or associated with rivers:

And when I turned my eyes to look beyond,
On a great river's bank I saw a throng ...
And lo! Toward us came, upon a barque,
An ancient man whose hair was long and white.[5]

Some may think that the *Divine Comedy* is just a poem, a work of imagination. But it is more than this. Dante's vision of the journeys of the dead was by no means a purely Christian or poetic notion. The various spheres of heaven and hell in the *Divine Comedy* are descended from a much earlier celestial geography. Most scholars conclude that Dante largely employed the cosmology of Virgil as it emerges in the *Georgics* and the *Aeneid*. Virgil, in turn, preserved the cosmic imagination of Plato as it appears in The Myth of Er in the *Republic*. Er was a soldier, slain in battle, who had a lengthy out-of-body journey in which the order of the spheres and the fate of the dead were revealed to him. He found himself at the gathering point for spirits entering and leaving the earth near a great column or 'Spindle of Necessity'. This column was in the middle of a celestial meadow and around it moved the planets. According to Plato:

Er, the son of Armenius, a native of Pamphylia . . . was killed in battle. When the dead were taken up for burial ten days later, his body alone was found undecayed. They carried him home, and two days afterwards were going to bury him,

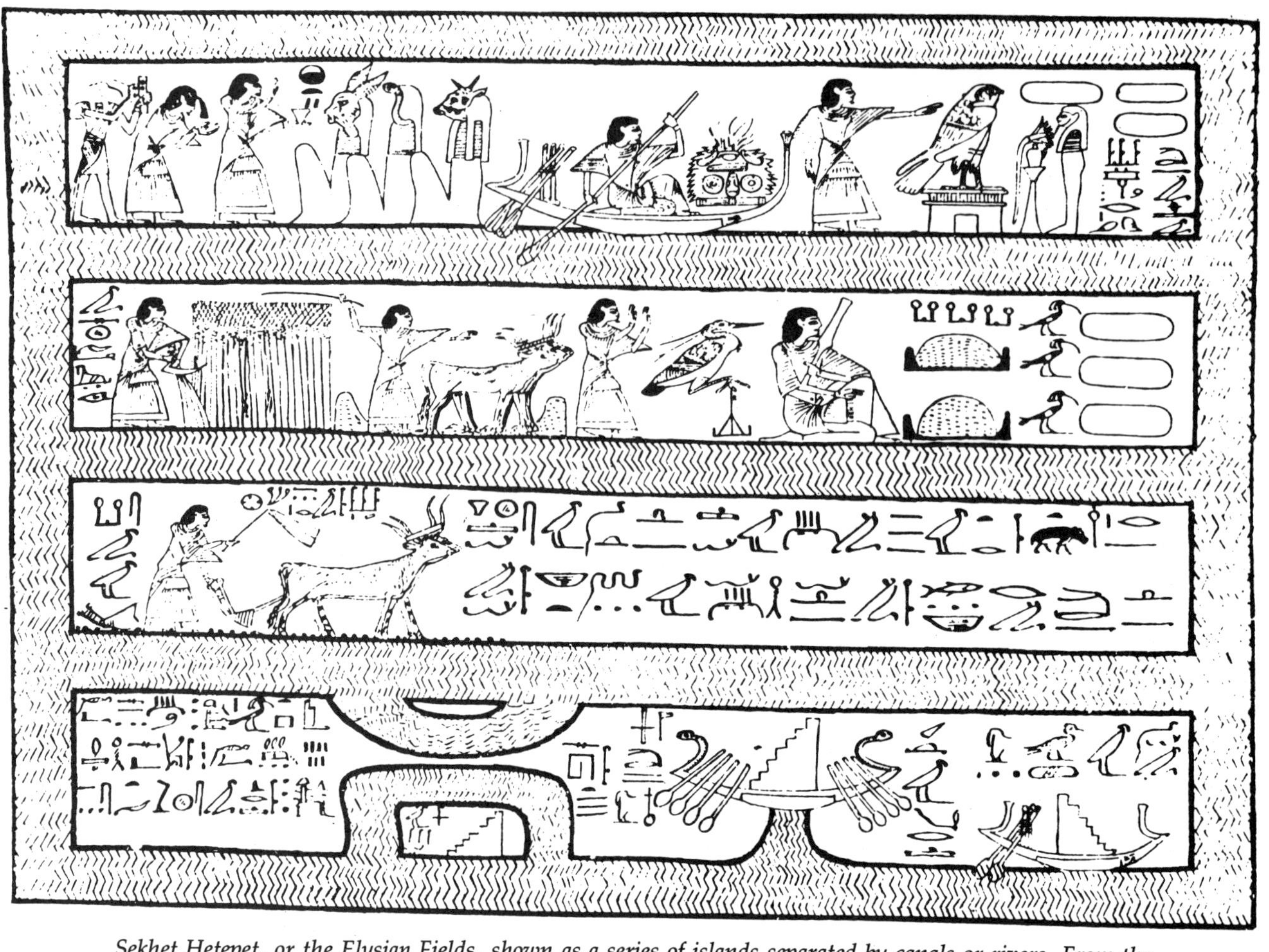

*Sekhet Hetepet, or the Elysian Fields, shown as a series of islands separated by canals or rivers. From the Papyrus of Ani in the British Museum.*

when he came to life again as he lay on the funeral pyre. He then told what he had seen in the other world . . .

They came to a place whence they could see a straight shaft of light, like a pillar, stretching from above throughout heaven and earth, more like the rainbow than anything else, but brighter and purer. To this they came after a day's journey, and there, at the middle of the light, they saw stretching from heaven the extremities of its chains; for this light binds the heavens, holding together all the revolving firmament, like the undergirths of a ship of war.

And from the extremities stretched the Spindle of Necessity, by means of which all the circles revolve. [Around this shaft or spindle] . . . there were in all eight whorls, set one within another, with their rims showing above us as circles and making up the continuous surface of a single whorl around the shaft, which pierces right through the centre of the eighth.[6]

Plato then numbers these eight whorls from the outermost in, describing the whorls as of varying breadth. Virtually all commentators on Plato are agreed that these eight concentric whorls are an image of the solar system. The outermost whorl — the fixed stars — and the shaft or spindle move slowly in one direction; the seven inner whorls—the planets—travel in the

*The ferryman of Dante's* Inferno *who carries souls across the first river of hell. The Greeks had an especially obstinate ferryman, Charon, who refused passsage to those unable to pay for his services.*

*The Egyptians, too, believed that one of the essential passages into the otherworld was across a river, which had to be navigated with the help of a celestial ferryman. The Egyptians called him 'Turn Face', since his boat points toward an outer shore while he faces backward toward the deceased.*

other. The spindle, another variety of the World Tree, is the sacred axis or column found throughout the earth.

Yet this imagination of the solar system did not begin with Plato; its origin is lost in the mists of antiquity. Plato had merely stood by the tradition passed on by Pythagoras (another initiate who lived in the sixth century BC) 'which called *cosmos* the order of the sun, moon and planets with what it comprised'.[7] Examining this cosmology in the context of similar mythic structures around the world, the eight whorls mentioned by Plato consistently appear as seven, eight or nine rivers. Even before Pythagoras, Hesiod in his *Theogony* described the world ocean Okeanos: 'with nine swirling streams he winds about the earth and the sea's wide bank ...'[8] The rivers of the Egyptian paradise also correspond to the planets: they are circles, seven in number and in them is found 'neither fish nor worm'. In the Mithraic mysteries the seven steps of the ladder of ascent were each associated with a planet. In Hindu and Buddhist cosmography there is a central mountain called Mt Meru which equates with the Axis, Tree, Ladder and Spindle. Again around this central hub there are seven concentric circles of oceans separated by seven intervening circles of golden mountains.[9] For Siberian shamans the world pillar was symbolized by the centre poles of their tents; these poles were customarily marked with seven notches, representing the seven heavens. The Aztecs had nine hells or purgatories bounded by rivers. The imagery is literally universal.

It must be emphasized that even when these are rivers of hell, they are *not* inside the earth, but out in space. The Lethe and Styx, for example, are rivers of Hades, but in *The Georgics* Virgil wrote: 'One pole is ever high above us, while the other, beneath our feet, is seen of black Styx and the shades infernal.'[10] De Santillana and von Dechend ask: 'What can it mean except that the Styx flows in sight of the other pole?'[11] And since these rivers of heaven are the destinations of the dead, and since they correspond to the circuits of the planets, what can it mean but that men travel to the planets after death?

In tracing the wanderings of the heroes of the *Aeneid* and the *Divine Comedy*, von Dechend says that 'the voyage really is not through subterranean caverns crowded with the countless dead, but through great stretches of emptiness suggesting night [and] space . . . '[12] Von Dechend further points out that 'Numenius of Apamea, an important exegete of Plato, comes out flatly with the contention that the other world rivers and Tartaros itself are the 'region of the planets'.[13] As the travellers in the Inferno proceed along the inner circle of the City of Perdition, 'there is a river of red boiling water'.[14] They later show that the 'circular territory occupied by the Red River in Hell was meant . . . to be the exact counterpart of the circle of Mars in the skies . . . '[15]

Many references to the region of the planets were obscured by the use of special terminology. In describing the domain of man, many ancient writers used the term 'earth' with a different meaning than ours: 'The true *earth* was nothing but the Pythagorean cosmos . . . '[16] In other words, it was the solar system.

From these and hundreds of related items of evidence, de Santillana and von Dechend also conclude that the rivers of heaven are planetary[17]— and then with their eyes fixed on the celestial content of other mythic elements, they pass by the extraordinary implications of the equation! According to the ancients, the rivers of heaven to which man travels after death are the circuits of the planets! What this means is that man — true man, immortal invisible man — is not merely a creature of the earth, but literally a cosmic being.

## Reincarnation and the Voyage Through the Planetary System

Like many archaic truths, the connection between the solar system and the after-death realm was never really lost. Medieval art from Europe to China still represented heaven with seven, eight or nine layers. Even today the expressions 'seventh heaven' and 'cloud nine' are very much alive.

Naturally, no single generalization can account for all the after-life myths of man or presume to describe the fate of all men, but the overall structure of the other world that emerges most frequently in the maps of consciousness available to us is that of the planetary system. The same vision appears in the *Upaṇishads*, the last part of the Vedic literature recorded a few centuries before Christ:

> Verily, when a person departs from this world he goes to the wind. It opens out there for him like the hole of a chariot-wheel. Through it he mounts higher. He goes to the sun. It opens out there for him like the hole of a drum. . . . He goes to the moon. . . . He goes to the world that is without sorrow, without snow. . . . [18]

In the final analysis, it is very difficult to explain how man could have acquired this celestial imagination unless he has had celestial experience. We are not dealing with a mere descriptive fantasy of a few ancient writers, but with a tradition of worldwide distribution and incalculable age. In uncovering man's larger domain, as in so many things, we in the late 20th century are merely rediscovering what the ancients knew long ago. In his essay 'On Exile' Plutarch quotes Empedocles and then continues with his own words:

> "It was not the mixture, O men, of blood and of breath that made the beginning and substance of your souls, though your earthborn and mortal body is framed of those things. But your soul has come hither from another place," and he speaks tenderly, in the gentlest of words, of its birth in a strange country.
>
> In truth the soul is an exile and a wanderer, driven from home by divine edicts and decrees, and then, as if in a sea-girt island, joined to a body, like an oyster to its shell, as Plato says. Hence it does not recall or remember the great honour and bliss from which it came. It exchanged, not Sardis for Athens, nor Corinth for Lemnos, but heaven and the moon for earth and life on earth.[19]

The number of sources, both early and late, that converge on this matter is truly amazing. In a rare instance of agreement, the two most famous mystics of the 20th century both declared that man travels to the planets after death. Rudolf Steiner (1861–1925) was a clairvoyant Swiss 'spiritual scientist', according to his own description. His hundreds of writings and recorded lectures on the origin and evolution of man are widely known and

*The Ninth and highest heaven of Dante's* Divine Comedy. *The geography of heaven as a series of spheres reflecting the circuits of the planets is a constant image over thousands of years and transcends cultural boundaries.*

*The spirits in Jupiter.*

*The moon.*

of continuing influence. The American trance-psychic, Edgar Cayce (1877-1945), a source of thousands of verified medical diagnoses, produced in addition hundreds of readings on metaphysical subjects, some of which are in areas also dealt with by Steiner.

The basic difference between Steiner and Cayce is that Steiner was a conscious clairvoyant, like Ingo Swann. Cayce, more like the shaman, went into a trance, losing consciousness yet speaking with his normal voice. Steiner 'focused his attention' on a subject; Cayce 'flew' to where the information was and reported back. Steiner and Cayce rarely describe things in the same way. Occasionally they seem to flatly contradict one another. Yet when it comes to the experience of man after death, they both repeatedly and consistently declared that it is to the planets that we go – including the sun and moon. They also agree that after we have visited the planets we return to earth and are born again into human experience.

Steiner sees the journey to the planets as a strict and orderly procession through all the spheres of the solar system. Cayce's itineraries are more dynamic, allowing the individual to visit the particular sphere or planet with the developmental influences he most requires, without having to take the whole tour before returning to earth. Cayce's descriptions of Saturn sound as if it is especially reserved for those who have erred greatly and are there 'recast' — a good approximation of hell, but residence on Saturn, or anywhere else, was not eternal.

According to Cayce, the ancients' knowledge of man's planetary sojourns was the key that led to the development of what we now call astrology. It was believed that the greatest planetary influence upon an individual was simply that of the planet from which he had just returned. Cayce spoke of an old Persian system of astrology that preserved that method of analysis — a system of which there is now little trace.

Of course, the journey to the planetary circuits is not a one way trip, otherwise we could hardly have heard about it. A significant element in initiation was not only to rehearse one's death, but also one's rebirth. In the *Aeneid*, Aeneas sees:

> Lethe's river, where it flows before the Homes of Peace. About this river, like bees in a meadow on a fine summer day settling on flowers of every kind . . . the souls of countless tribes and nations were flitting.

His father Anchises tells him:

> They are souls who are destined to live in the body a second time, and at Lethe's wave they are drinking the waters which abolish care and give enduring release from memory.[20]

Before Virgil's Aeneas, Plato's Er said much the same thing about the River of Unmindfulness, but implied that the wisest drank the least. For Plato, rebirth was a necessity until one was sufficiently wise and pure to leave the solar system entirely. Er describes heroes and demigods choosing their next lives on earth and has crafty Odysseus make the best choice—'a life of quiet obscurity'.

The belief in reincarnation is extremely ancient and widespread. It occurs not only in Platonic and Pythagorean philosophy but in Buddhism, Hinduism, Jainism, in the Persian Mithraic and Greek mystery religions, among the ancient Celts, the Eskimo, various American, Polynesian and Asian peoples, and in early Christianity. In *De Principiis*, the Church Father Origen stated:

*Sunset at the Great Pyramid.*

*The Third Gizeh Pyramid with subsidiary pyramids to the left.*

The soul has neither beginning nor end.... Every soul comes to this world strengthened by the victories or weakened by the defeats of its previous life. Its place in this world as a vessel appointed to honour or dishonour is determined by its previous merits or demerits. Its work in this world determines its place in the world which is to follow this.

In AD 543 this early and widespread doctrine was attacked by the Byzantine emperor Justinian and later condemned as heresy by the Second Council of Constantinople in 553. The Council seems to have also condemned the disciples of Jesus who asked about a man blind from his birth: 'Who sinned, this man or his parents, that he was born blind?' (JOHN 9:2). Other references to reincarnation can be found in the Bible[21] and many recent Western studies supporting it can be cited.[22]

There is much about death the modern world might learn from ancient and archaic peoples. The equanimity and poise, the calm acceptance of death and even a disdain of death was a distinguishing characteristic of many ancient peoples. Knowing that death was but another birth and not a final end, they regarded themselves as eternal. The Egyptians especially were notorious for this. It gave them the confidence to build great structures that would endure for hundreds of generations; it breathed into their entire civilization a spirit of grandeur, calm and confidence which reaches us still in the testimony of its great architectural remains. Among the ancient Celts the utter confidence in rebirth of the spirit in another body led Caesar to complain in *De Bello Gallico* that their warriors would appear in battle naked (giving them great quickness) and engaged in war with such vehemence as to show complete disdain of their own safety. 'Hence their warrior's heart hurls them against the steel, hence their ready welcome of death, and the thought that it were a coward's part to grudge a life sure of its return.'[23]

Most of us do not remember our previous lives since, as the Greeks put it, in passing between the worlds we drink the waters of forgetfulness. These waters, however, do not affect everyone equally. The wisest drink only a little. Pythagoras reportedly could remember as many as ten lives. A Winnebago Indian shaman called Thunder-Cloud remembered three:

It is I, Thunder-Cloud, who speaks, I who am now on earth for the third time, I who am now repeating experiences that I well remember from my previous existences.

Once, many, many years ago I lived with people who had twenty camps. When I was but a young boy . . . our village was attacked by an enemy war party and we were all killed.... All I knew and felt was that I was running about just as I had been accustomed to do. Not until I saw a heap of bodies on the ground, mine among them, did I realize that I was really dead. No one was present to bury us. There on the ground we lay and rotted.

Then I was taken to the place where the sun sets and I lived happily for some time with an old couple. In this particular village of spiritland the inhabitants have the best of times. . . . after a few years, the thought ran through my mind that I would like to return to earth. . . .

So I went to the chief and I told him my desires. . . . "Go, my son, and obtain your full revenge upon those who killed all your relatives and you."

Thus was I brought back to earth again. . . .

I was being born from a woman's womb. . . .

I was taught to fast in order to prepare myself adequately and completely for warfare, for my purpose in returning to this earth. In due course of time I went on the war path. I did so repeatedly until I had taken full revenge for the death of myself and my dear relatives. . . .

I finally died of old age. . . .

This was the second time that I had known death. . . .

There in the grave I lay and rotted.

As I was lying there, rotting, I heard someone speak to me saying, "Come, we must leave now!" Obediently I arose and we walked, the two of us, in the direction of the sun where lies the village of the dead. There all the people are gay and enjoy themselves. . . . Four nights I was told I would have to stay in the village, but they meant four years.

When my stay was over I was taken up to the home of Earthmaker. There at his own home, I saw him and talked to him, face to face, even as I now speak to you. The spirits, too, I saw. In fact I was one of them. Then I came back again to this world. Here I am.[24]

CHAPTER TEN

# THE GOLDEN PLATE OF PETELIA

## Face to Face with Cosmologies Past and Present

The stratagems of time have served us differently than we thought. The more the modern world unravels the beliefs of ancient peoples and explores the frontiers of man's being, the more it appears that we are coming full circle into an area of extraordinary synthesis in which the intellectual divisions separating the past and present are suddenly shattered by the shock of recognition.

Thunder-Cloud and Pythagoras would have understood one another; Black Elk and Plato shared the same opinions. In his famous allegory of the cave, Plato likens this world to a subterranean chamber that can be reached only through a long passage to the upper world. Higher up is a fire that throws shadows on the walls of the chamber. This chamber of shadows merely reflects events in higher realms above this one, an image which has become synonymous with Platonism itself.[1]

It is startling to find that Black Elk, who certainly had never read Plato, described terrestrial experience in the same way and with the same metaphor. Speaking of the vision of Crazy Horse, his well-known cousin, Black Elk says:

*The gold chain and case which held the plate found at Petelia.*

Of course [my father] did not know all of it; but he said that Crazy Horse dreamed and went into the world where there is nothing but the spirits of all things. That is the real world that is behind this one, and everything we see here is something like a shadow from that world.[2]

Accustomed to seeing only the outward variations in cultures and religions, at first it seems astonishing that the cosmologies of the Buddhists, Tibetans, Egyptians and of such different men as Plato and Black Elk all come to much the same thing. And, initially, it seems almost incredible that these cosmologies can actually be used to elucidate the clinical findings of contemporary physicists, psychologists and physicians dealing with death and out-of-body flights. But if there is any truth in the old cosmologies, and if it is indeed all one world, then what is surprising is not the convergence of ancient cosmologies, but the divergence of our own.

In exploring the world charted on the star maps, we are learning how much they have to teach us, and how profoundly their vision of man differs from current opinions. Contemporary anthropology sees man as a creature evolved from earth but if, as Empedocles says, our souls have 'come hither from another place' then we must ask again, from where did we come? What kind of being is man?

If our involvement with the circuits of the planets depends upon our deeds on earth, and our travels to and fro are part of a continuous process in which the sun, moon, earth and planets are one system — if the true earth is the entire solar system — then it seems that originally man came from the stars.

Very much as the Egyptians placed chapters of the *Book of the Dead* in the casket with the deceased, during the fourth to the second centuries BC Orphic initiates in southern Italy and Crete carried written instructions on how to navigate in the other world with them into the grave. These instructions were impressed into thin rectangular sheets or plates of gold, about 5 x 7.5 cm (two inches by three). Seven such plates, some fragmentary, have been found. The oldest, a plate from Petelia, southern Italy, was rolled up, not without some creases, and inserted in a small cylinder suspended on a 27.5 cm (11 inch) chain with a hook at one end and an eye at the other. It both adorned and protected its owner, serving as an amulet.

In these brief fragments recorded in gold are key phrases spelling out man's nature and origin in language as cosmic as it is poetic. They also contain the reason why Unas aimed for the circumpolar stars.

Thou shalt find to the left of the House of Hades a spring,
And by the side thereof standing a white cypress.

To this spring approach not near.
But thou shalt find another, from the Lake of Memory
Cold water flowing forth, and there are guardians before it.
Say, "I am a child of Earth and starry Heaven;
But my race is of heaven alone. This ye know yourselves.
But I am parched with thirst and I perish. Give me quickly
The cold water flowing forth from the Lake of Memory."
And of themselves they will give thee to drink of the holy spring,
And thereafter among the other heroes thou shalt have lordship.[3]

That is the entire text of the Petelia plate. Another, from Eleuthernai in Crete, about two centuries later, is even shorter and more fragmentary, but just enough remains to see that the owners of these plates craved the same drink and claimed the same origin:

I am parched with thirst and I perish—
But give me to drink of
The ever-flowing spring on the right, where the cypress is.

The bearer is asked:

Who art thou?
Whence art thou?

And he answers:

I am the son of Earth and starry Heaven.[4]

The cold water from the Lake of Memory and the everflowing spring on the right remind us again of the Cool Region and the purification of Unas in the waters of the circumpolar stars—the stars to which the pyramid passages point. They also sound like the heavenly waters in Dante's Paradise:

Whoso lament that one must die below
To gain a heavenly life, has not seen here
The sweet refreshment of the eternal rain.[5]

In the Myth of Er, Plato too distinguishes between those on the right and those on the left. Er witnesses a judgment scene in which those on the left are sent 'downward' back to earth and the planets, while those on the right are sent 'upward' just like Unas. Of course, the same left-right distinction is found in the Bible; the goats on the one hand, the sheep on the other, those found lacking, and the redeemed. Speaking of the Day of Doom in the Koran, Muhammad separates the Companions of the Right from the Companions of the Left, and declares them to be the blessed and the damned, respectively.[6]

*The gold plate from Petelia, southern Italy, third or fourth century BC.*

Muhammad also speaks of Outstrippers. 'Those are they brought nigh the Throne', he tells us; [7] it is easy to see that they are the same sort of people as Black Elk, the 'saints' and higher initiates who, like Unas, 'become being'. This evolution to a higher condition is spelt out in another Orphic funerary plate from Thurii, southern Italy:

But as soon as the spirit hath left the light of the sun,
Go to the right as far as one should go,
Being right wary in all things.
Hail, thou who hast suffered the suffering.
This thou hadst never suffered before.
Thou art become god from man.
A kid thou art fallen into milk.
Hail, hail to thee journeying the right hand road
By the holy meadows and groves of Persephone.[8]

In virtually all ancient cosmologies the west was the land of the dead. From the Egyptians at the time of Unas to the Greeks, Celts and peoples of western Europe, to Black Elk, it was believed that at death the soul first travelled west, following the setting or dying sun. Using the western sun as the point of orientation, turning to the right 'as far as one should go' would again send the spirit north and to the circumpolar stars; turning left would bring it to the opposite region in the south — the location of the Underworld and the river of Lethe. The right hand road leads to the primal

*Like Eleusinian initiates thousands of years later, Egyptians sought the celestial waters springing from the top of the cosmic tree. The water sustained both body and spirit (note the soul drinking) and those on both levels of existence (heaven and earth). In this case the water is dispensed through an intermediary, a goddess, who also offers food. On either side of the trunk stands a Bennu bird, symbol of immortality. The tree rises from a realm of fish and fowl, signifying earth and its depths.*

water of life. As milk is to the life of a kid, so is this water to the soul of man. Those wise and strong enough to attain it are transformed: 'Thou art become god from man.'

Finally, in an inscription reconstructed from three fragmentary plates also found near Thurii, we have the clearest statement of the reason why thousands of years earlier Unas went to the circumpolar stars:

I come from the pure, pure Queen of those below,
And Eukles and Eubuleus, and other Gods and Daemons:
For I also avow that I am of your blessed race.
And I have paid the penalty for deeds unrighteous,
Whether it be that Fate laid me low or the gods immortal
Or (that I was stung) with star-flung thunderbolt.
I have flown out of the sorrowful, weary circle.
I have passed with swift feet to the diadem desired.
I have sunk beneath the bosom of the Mistress, the Queen of the Underworld.
And now I come a suppliant to holy Persephoneia,
That of her grace she send men to the seats of the Hallowed.
Happy and blessed one, thou shalt be god instead of mortal.
A kid I have fallen into milk.[9]

'I have flown out of the sorrowful, weary circle.' This circle is nothing but the plane of the planets and the cycle of rebirth. If man has paid his debts for deeds unrighteous in the Underworld and wishes to avoid another round of existence in planetary realms, he must aim away from them and for that place specified again and again in archaic maps of consciousness as the haven of the fully evolved man. At first the planetary realms may seem fascinating to the earthbound, but as one grows in experience and comes to know the entire system, after great ages he will become weary of even the rivers of heaven.

It is repeatedly stated in the Text of Unas that he had become 'weary of the Nine', and then it is indeed time to fly out of 'the sorrowful, weary circle'. Unas went to the circumpolar stars because in rehearsing his death he wished to avoid further involvement with the solar system and the Underworld entirely. He had 'graduated', and though 'a child of Earth and starry Heaven', henceforward his race would be of heaven alone—as it had been in the beginning.

This destiny in the outer heavens is clearly stated in the Dream of Scipio, the conclusion of Cicero's treatise *De Republica*, probably written around 54 BC. Scipio relates a dream in which he left his body and spoke with his grandfather, Scipio Africanus the Elder. For this patriotic Roman, a life lived in service to his country was his 'highway to the skies'. His grandfather urges Scipio, who wishes to remain in heaven, to return to earth and to live a life of service, so that he may join:

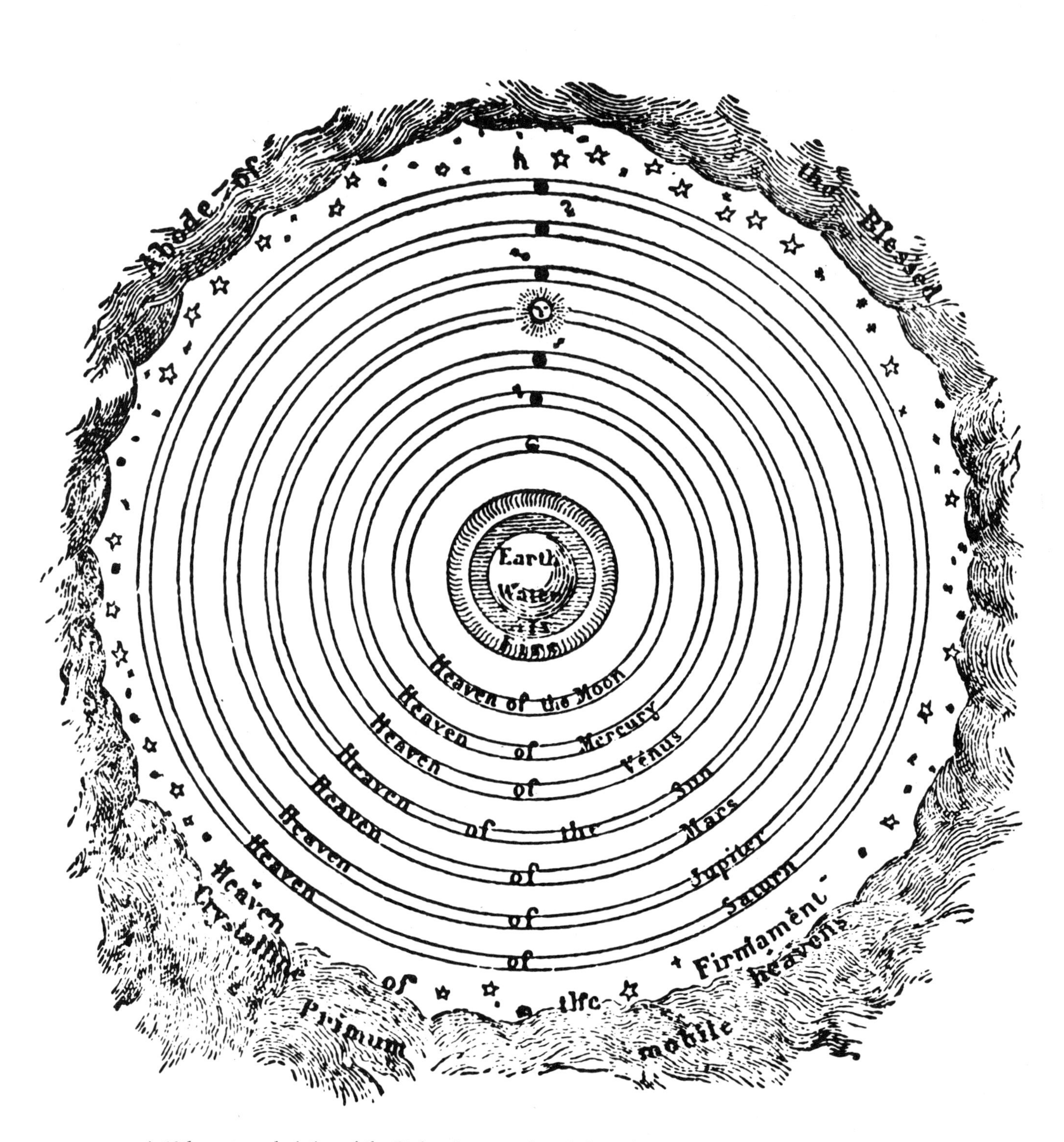


*A 19th century depiction of the Ptolemaic conception of the universe with the earth at the centre, and illustrating the ancient tradition that the Abode of the Blessed lies outside the circuits of the planets.*

. . . the company of those who, after having lived on earth and escaped from the body, inhabit the place which you now behold.

This was the shining circle, or zone, whose remarkable brightness distinguishes it among the constellations, and which, after the Greeks, you call the Milky Way.

From thence, as I took a view of the universe, everything appeared beautiful and admirable: for there, those stars are to be seen that are never visible from our globe, and everything appears of such magnitude as we could not have imagined. . . . The globes of the stars far surpass the magnitude of our earth, which at that distance appeared so exceedingly small, that I could not but be sensibly affected on seeing our whole empire no larger than if we touched the earth as it were at a single point.[10]

## The Extraterrestrial Nature of Man's Origins

In following the trail of Unas and his kind, we finally come to the edge of an enormous chasm, the bottom of which cannot be seen, and beyond which there are only the stars. From them we came, an everlasting race of starry heaven, sent to the earth and its environs, to 'subdue' it and the bodies we acquired there.

The maps of consciousness we have surveyed and the star maps we have explored are all maps of the same country. The story of man's origin, history and status in the universe as it emerges from these charts is indeed so fantastic that beside it the most colourful science fiction pales in comparison. Man is an ancient, ancient being. We have passed through many experiences now so long ago and far away that we have forgotten them almost entirely—until memory rises from deep unconsciousness.

If the history of archaeology illustrates nothing else, it shows that there is often an enormous gap between a physical discovery and the perception of its significance. The old maps analyzed by Hapgood were known to scholars for literally hundreds of years before he perceived that they record information that could not have been known to Renaissance cartographers. Many of the ancient American stones inscribed with Mediterranean languages recently decoded by epigraphers Fell, Gordon and Mertz were known for over a century before their significance was perceived. Similarly, Stonehenge and the Great Pyramid fuelled speculations for millennia before recent discoveries have allowed us to see them in a more sophisticated light.

And so it is with what might be called the ultimate discovery of archaeology: the incredible evidence that man himself is an extraterrestrial has been available for decades. It is only very recently that laboratory science has begun to probe the very aspects of man that were so highly developed by the ancients. When the implications of this research are applied to enigmas like the Pyramid Texts and the curious worldwide dispersion of certain aspects of initiation, we have a conceptual scaffolding whereby anyone who is willing to climb can reach the perspective to see what poets and mystics have always known: we come from the stars.

Initiatory journeys did more than acquaint one with the realms beyond and exalt his spiritual nature; they also recapitulated the history of man. As Eliade has remarked:

> It is not possible here to pass in review all the species and varieties of this "flight" and of the communications between Earth and Heaven. Let it suffice to say that the motif is of universal distribution, and is integral to a whole group of myths concerned both with the celestial origin of the first human beings and with the paradisiacal situation during the primordial *illud tempus* when Heaven was very near to Earth and the mythical Ancestor could attain to it easily enough by climbing a mountain, a tree or a creeper. . . . In short, the ascension and the "flight" belong to an experience common to all primitive humanity.[11]

It is precisely the fact that myths describing the celestial origin of the first human beings are of universal distribution that makes it virtually impossible to account for man as anything but an extraterrestrial. Of course, some historians suggest that ancient man was universally deluded, but even if we entertain this possibility, it is extremely difficult to explain why peoples literally all over the earth had the *same* delusion. Myths the world over proclaim man's origin outside and beyond the earth. Among the Pawnees of Nebraska, Tirawa the great chief in heaven told the sun and moon to unite, to people the Earthly Paradise, and a son was born. Then the morning and the evening stars united, producing a daughter. These two children were placed on earth and when they had grown up, Tirawa sent gods to teach them the secrets of nature. Other men were created by the stars.

Among the Iroquois, the peopling of earth began with the uprooting of the Tree of Life in heaven. A goddess was thrown into the hole left by the Tree. She fell to earth and was mother of the gods of creation who made men and women. In Hawaii, Northern Australia and New Guinea the ancestors of men were gods who came down to earth from heaven. Much of Japanese mythology is structured to trace the celestial origin of the Japanese royal family. Formerly, earth was linked with heaven by a floating bridge which allowed the gods to go to and fro. One day when the gods were all asleep this bridge or stairway collapsed into the sea, stranding some on earth.

Most mythologies distinguish the creation of the physical body as a separate and last act. Typically, this body was formed from the most prosaic of materials: dust, clay, grass, wood, maize and even excrement. The spirit that comes into it is another thing, preexisting and infused by the gods through incantations, breathings, elixirs, and according to the Narrinyeri of South Australia, through laughter. The creator tickled the forms he made to make them laugh and give them life.

The motif is ever the same: man came into materiality from outside; his spirit is not native to the earth. This is again discernible in the varying accounts of the forms of man before he took on flesh. According to Plato in the *Timaeus*, the first men were circular androgynes (having both sexes in the same body) with four arms and four legs. They were so powerful that they threatened the gods who decided to split them in two, so that one half would always be looking for the other (its soul mate) and would trouble the gods no more. This closely parallels the description of the natives of the Society Islands in the southwest Pacific: men did not at first have human shape, but were like balls on which arms and legs developed later. Similarly, among Australian tribes, the first men were said to be of a rounded shape with only rudimentary appendages and organs which were later developed by deities or supernatural beings.

The celestial origin and nature of man is most strongly emphasized in the symbolism applied to man's soul. Birds, bees and butterflies are metaphors for the soul which has not ascended. However, souls which have escaped the sorrowful, weary circle, and souls as they were before they descended to earth are ever likened to stars. 'Unas appears as a star', it reads in the text on the walls of his pyramid. The same idea applied to other high initiates runs throughout the pyramid texts. According to Plato, before their involvement in this system, each soul came from a different star, and would return to it after graduation from the Nine. The connection is carried further by Cicero: 'Minds have been given to [men] out of the eternal fires you call fixed stars and planets, those spherical bodies which, quickened with divine minds, journey through their circuits and orbits with amazing speed.'[12]

Again the same symbolism is found around the earth. Among the Maya 'there is a clear notion that when a person dies he becomes a star. The better the person, the bigger the star. . . . When a baby is born it is one of the stars in the sky come back to earth as the soul of a person.'[13] Like Cicero, the Chinese and the Hopi Indians saw the Milky Way as the path of the souls to the outer heavens.

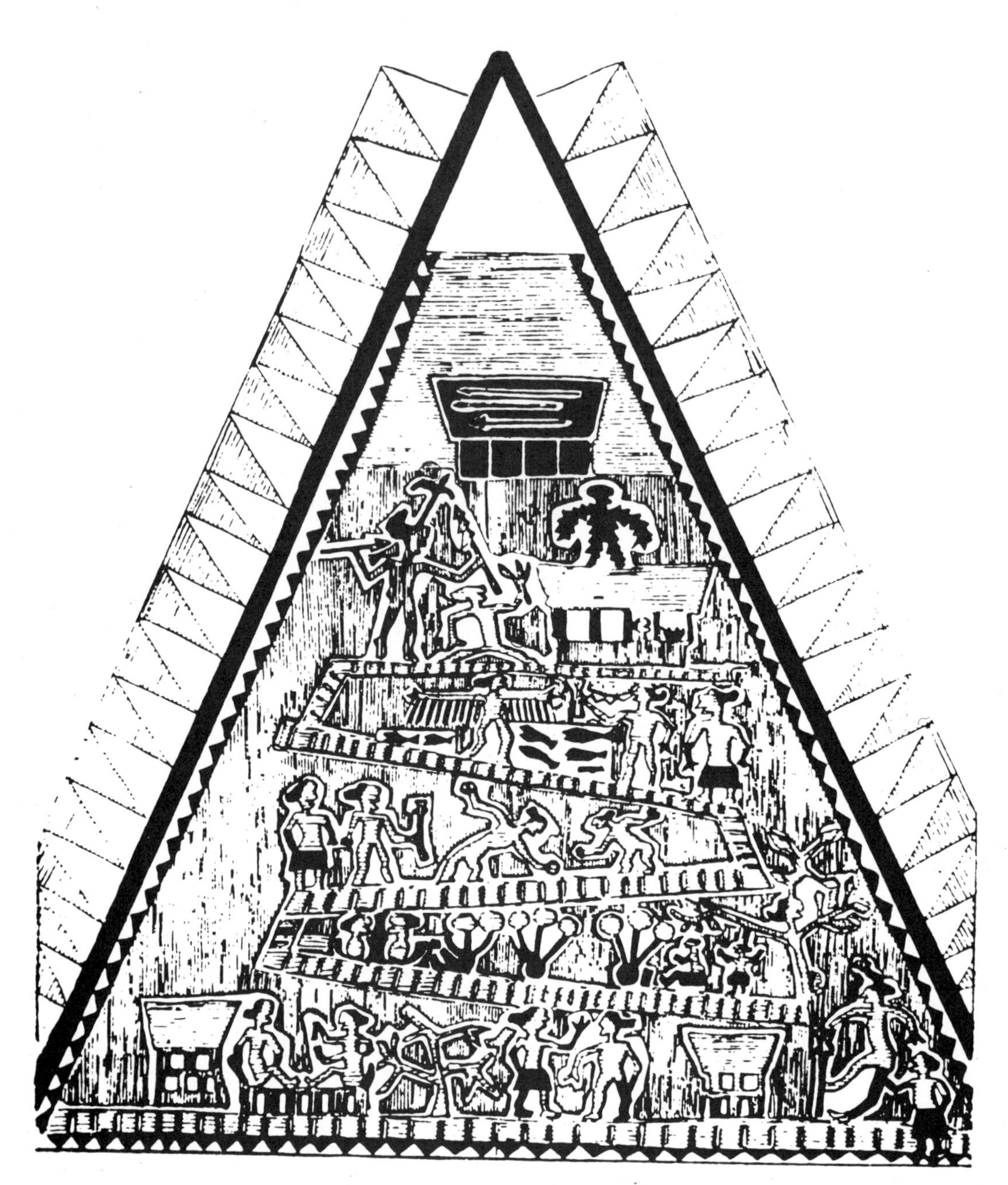

*The ascent of spirits of slain heroes wandering over The White Way into heaven. From a gable painting in a men's long house in the Palua Islands of the western Pacific.*

The soul in its pristine state is often described as a star. The 16th century Galileean cabbalist, Isaac Luria, likened the descent of souls into the earth as a shower of 'sparks of light'.[14] This description conjures up images from Doré: billions of tiny stars flowing like a river of light through the far reaches of space come upon the blue and golden earth; they descend; they become trapped in matter and ever after long for their celestial home.

Man's celestial origin is betrayed in his very behaviour. From the oldest text on earth until the present, an instinctive fascination with the stars has

*Virgil and Dante behold the stars.*

*A 19th century Chucki drawing from Eastern Siberia, showing various celestial bodies, the Milky Way and the connection between worlds.*

linked men everywhere. Surely if an educated Egyptian of 5000 years ago were somehow transplanted into the fourth quarter of the 20th century, he would conclude that we are just as star-struck as the pyramid builders. All over the earth giant telescopes peer deep into space photographing and numbering the galaxies. Untold sums of money are spent on moon landings, space probes and orbiting satellites in enormous programmes comparable in terms of national efforts to the building of the pyramids themselves. Plaques and signals are sent toward the stars in government-funded projects attempting to communicate with other beings. Millions are convinced that UFOs are beings from other solar systems attempting to communicate with us. Imaginative adventures in the stars take place in thousands of literary works of science fiction, in motion pictures, television, poems and songs. Millions more believe that the positions of the stars influence their destinies for good or ill.

A 5000 year old Egyptian would conclude that with the great space programmes of the 20th century we are merely attempting to accomplish with material bodies and instruments what the ancients accomplished with immaterial bodies and a knowledge of man's true nature and origin. He would report that with our narrowly focused science and philosophies, we have forgotten the great primary truths about ourselves. We are still reaching for the stars, even as Unas reached for them over 40 centuries ago, but we are just rediscovering that our star ships are within us. Money spent on the exploration of space will surely accomplish something, but it would

*Myths describing the celestial origin of the first human beings are of universal distribution. Our own galaxy is thought to look much like the great spiral galaxy in Andromeda (M31), estimated to contain 100 billion stars.*

seem that Unas travelled further than any modern astronaut and that his space programme was far more successful than ours. To the wise men of Greece, Egypt, Crete, Tibet and the American West, our rockets and satellites would seem nothing more than toys performing the same function for us as the playthings of our children do for them.

CHAPTER ELEVEN

# THE LAKE OF MEMORY

## The Universal Path of Initiation

All over the earth the aim of initiation was the separation of the spiritual from the physical so that man would not become too wedded to the earth, so that he would not lose his way among the stars. The day of the old initiations has passed, and we in this most materialistic of centuries may feel ourselves forever estranged from that age of grace and mystery. But the past is not dead; it lives in each of us.

The perceptions of the ill and injured who cross over into the other realm and return to tell the tale are generally patterned after the expectations to which they have been educated. But not always. Grof and Halifax point out that in these experiences:

> European and American subjects do not necessarily follow the canonic rules of the Judaeo-Christian religious tradition, as one would expect. On occasion unsophisticated individuals have described detailed sequences from Hindu, Buddhist, and Jain mythology, or complex scenes from the little-known Egyptian *Book of the Dead*. . . . In many instances, the sophisticated structure of such sequences transcends the education of the experiencer.[1]

It may be that thousands of years ago the sacred initiations, mysteries and knowledge were the province of the few, but through long ages of life after

*The Rosette Nebula.*

life everyone undergoes the travels of the initiates. Consciousness accumulates from life to life, filling the lake of memory, until one day the lake overflows. What must it do to a good Presbyterian to wake up and find himself towing the Great Barque of Ra? What would it do to *anyone*, in dream or waking vision, to see with unmistakable clarity the end of his previous life on earth?

Each man is on the path of initiation by his very involvement in the cycles of life, death and rebirth. However slowly, he, too, may find the cosmic centre that is the gate to the stars. The great Tree, Pillar, Mountain or Ladder by which the shaman made his ascent exists on many levels. On the grand scale it is the celestial axis of the solar system. On the family or tribal level, this same *axis mundi* was symbolized by the centre pole of the tent. Among peoples who constructed tepees with a hearth in the centre and a smoke hole at the top, this smoke hole became the entrance to the other world. On the individual level the Tree, Pillar, Mountain or Ladder is represented by the spine, 'the centre pole' of the physical body. It was for this reason that the Djed column represented both the world axis and the backbone of Osiris.

The cosmic centre is anywhere and everywhere. When he was called up by the Grandfathers, Black Elk said the mountain 'at the centre of the world' to which he was taken was Harney Peak in the Black Hills, but carefully added that 'anywhere is the centre of the world'.[2] Perhaps the earth was depicted at the centre of ancient maps of the cosmos because the ancients shared Black Elk's perspective. The earth was so placed, not through ignorance about the structure of the solar system, but because a map is most convenient when one's location is at the centre—and this is especially true of a map of consciousness for an observer on a spinning globe who sees the planets and stars moving around him. It is clear from many ancient sources that the ancients were well aware of the heliocentric circuits of the earth and the planets, but in later classical times and during the medieval era, the heliocentric concept was lost and the meaning of the cosmic centre was subsequently misinterpreted.

Since the centre of the world is everywhere, each of us has his own centre. Throughout the world, the religions and cosmologies of man concur that at his uttermost core there is a light that is the heart of everyone. In the Far East, initiation often consisted of finding that light within through meditation. Each man contains in substance and pattern a part of everything in creation. The seven spiritual centres of the body (associated in Western metaphysics with the endocrine system), or the seven *chakras* as they are known in the East, correspond to the planets. If one can focus his attention and vital force in his highest centre, he too can fly out of the

*The Whirlpool Galaxy. The further man looks within, the more he sees visions comparable to the lights of the universe.*

sorrowful, weary circle. As the Chinese sage Lao Tsu, author of the *Tao Te Ching*, put it in the sixth century BC, 'without going outside, you may know the whole world. Without looking through the window, you may see the way of heaven.'

Again, this is not a matter of poetry, but of experiential reality. Centring on the light within produces much the same state as that attained by Unas: 'Thou hast become being . . . thou art become spirit.' As John Lilly discovered, it is not necessary to leave the body to reach this state:

There are multiple planes of being . . . intersecting at very high energy through my body's vertical axis. I am in Zen kneeling position. . . . If I keep the line vertical . . . I stay centred. . . . I hold on to the line. Amazing energy sweeps through me. . . . When I accept the laws, I am in divine cosmic love. I can take the energy, and stay centred on the line. Immediately after this, I experienced a liquid gold-red light pouring out of the cosmos onto and down through me, with immense love and gratitude flowing around every cell of me. I become illuminated and enlightened and immensely happy.[3]

Another out-of-body traveller, Elizabeth Kübler-Ross, also found she could reach the same state in the flesh.

I had a vision of the whole universe, everything vibrating. And in front of me something opened. . . . The moment that I focused on it, it turned into a lotus flower bud. It had the most incredible colours — beauty that I cannot put into words. . . . Behind this flower bud came something like a sunrise — an incredible light. . . . I looked at all this in utter awe. There were sounds and colours and visions beyond description.[4]

It is much the same cosmic vision reported throughout time. In the 16th century, 25-year-old Jakob Boehme found himself 'surrounded by divine light, and replenished with the heavenly knowledge' as he wandered in fields near Görlitz. Of a later experience he wrote:

In one quarter of an hour I saw and knew more than if I had been many years together at a university. For I saw and knew the being of all things. . . . I saw and knew the origin and descent of this world, and of all creatures through divine wisdom.[5]

Robert Monroe reached the same state out of the body.

Three times I have "gone" to a place that I cannot find words to describe accurately. . . . It is this vision . . . or state of being that brings the message we have heard so often throughout the history of man. . . . This may be part of the ultimate heaven as our religions conceive it. It must also be the nirvana, the Samadhi, the supreme experience related to us by the mystics of the ages. It is truly a state of being, very likely interpreted by the individual in many different ways. To me it was a place or condition of pure peace, yet exquisite emotion. . . . Nothing exists

*The seven* chakras.

as a separate piece of matter. The warmth is not merely around you, it is of you and through you. Your perception is dazzled by the Perfect Environment. . . . This is where you belong. It is home.[6]

Monroe speaks of his return in terms that call to mind the words of Plutarch in his essay 'On Exile'. Monroe did not leave this 'home' of his own will but was led back, and felt himself an alien on earth.

In sketching the travels of man beyond this earth, the planets may be the territories, the crystallizations of particular dimensions, as it were, to which we journey; but the scenes to which one gravitates within these realms have more to do with longings of the heart than with the geography of space. In the Egyptian judgment scene, it is the heart that is weighed. With their literal manner, the Nile dwellers often put a small stone heart amulet under the tongue of the dead. In *The Teachings of Don Juan* an old Yaqui Indian sorcerer experienced in seeing more than just the physical dimension, instructs his pupil to 'follow the path that has a heart'. It was by following Beatrice, the love of his heart, that Dante trod the paths of heaven. Many ancient traditions concur on the remarkable opinion that it is the heart that actually thinks. We all know a special power when we put our heart into something. Regardless of what we may or may not outwardly profess, it is what we have in our heart that determines what we experience after death.

*Dante following Beatrice to the circuits of the heavens.*

Among the most intriguing of George Ritchie's visions on the other side was an enormous university. It was, he says, 'as if all the schools and colleges in the world were only piecemeal reproductions of this reality'.[7] The people inside were garbed in loose-flowing robes like monks but it was not a monastery. Through open doors Ritchie glimpsed huge rooms full of complex equipment. It was, he recounts, like a 'tremendous study centre', and 'whatever else these people might be, they appeared utterly and supremely self-forgetful — absorbed in some vast purpose beyond themselves'.[8]

What he saw was literally a scholar's paradise. If our heart is in the pursuit of knowledge, we shall find it; if with family or loved ones, we shall find them; and if with the light at our centre, we shall find that too. If these many visions of peace and love and light are not mere fantasies but experiential realities, it is more sobering to consider that other visions might have the same basis, such as this from Dante's Inferno:

Vengeance of God ! What dread must thou inspire
In everyone who now shall read and learn
Of that which was revealed before my eyes!
For I saw multitudes of naked souls
Who all were weeping piteously, and seemed
Tormented all in varying degrees.
While some were lying supine on the ground,
Others were sitting huddled in a heap,
Or running about incessantly.[9]

Many in the modern world may believe that we have somehow outgrown such images, but the compilation of after death experiences in recent years suggests otherwise. Guided by a being of light, who Ritchie believed was Christ, the Virginia doctor's travels in that realm seem more extensive than most other current accounts. As well as scenes of grandeur and light, he was also shown the fate of the self-righteous, the perverse and the violent:

The plain was crowded, even jammed with hordes of ghostly discarnate beings. . . .

At first I thought we were looking at some great battlefield: everywhere people were locked in what looked like fights to the death, writhing, punching, gouging. It couldn't be a present day war because there were no tanks or guns. No weapons of any sort, I saw as I looked closer, only bare hands and feet and teeth. And then I noticed that no one was apparently being injured. There was no blood, no bodies strewed upon the ground; a blow that ought to have eliminated an opponent would leave him exactly as before. . . . They could not kill, though they clearly wanted to, because their intended victims were already dead, and so they hurled themselves at each other in a frenzy of impotent rage.

*Torment in hell visualized by Gustave Doré.*

*Strife in hell. Doré's illustration is remarkably close to Ritchie's out-of-body description.*

*Dante's vision of the Forest of Suicides. Interestingly, Moody tells of an attempted suicide who reportedly found herself enmeshed or trapped in the very situation from which she had tried to escape.*

> If I suspected before that I was seeing hell, now I was sure of it. . . . These creatures seemed locked into habits of mind and emotion, into hatred, lust, destructive thought-patterns. . . . Even more hideous than the bites and kicks they exchanged were the sexual abuses many were performing in feverish pantomime. . . .
>
> And the thoughts most frequently communicated had to do with the superior knowledge, or abilities, or background of the thinker. "I told you so!" "I always knew!" "Didn't I warn you!" were shrieked into the echoing air over and over. . . . [10]

Perhaps the images of the other realm pictured by Dante and by the famous illustrator, Gustave Doré, are more real than we think. Surely on the islands of the rivers of heaven there is sufficient room for many mansions and, for all our supposed enlightenment, there may indeed be more in both heaven and earth than has been dreamt of in the philosophies of our time.

*Harpies in the Forest of Suicides.*

CHAPTER TWELVE

# THE GODS OF LIGHT

## Consciousness Before Life on Earth

Where is the way to the dwelling of light,
and where is the place of darkness,
that you may take it to its territory
and that you may discern the paths to its home?
You know, for you were born then,
and the number of your days is great![1]

So God answered Job out of the whirlwind. The message of the star maps and the many testimonies of antiquity is that man is older than the oldest bone on earth. In the ancient vision, man preexisted the development of the physical body he came to inhabit, a body which for him is just a garment, a skin to clothe his invisible being and to provide a temporary domicile for his still unfolding consciousness. Man was a conscious sentient being *before* he came to earth. Perhaps, to use the image suggested earlier, he appeared like hundreds of millions of tiny stars that had comprised a living river of light which flowed from an unnameable source. Man came freely of his own will and choice, descending *en masse*, like a shower of sparks of light, as Luria says, suddenly and at once all over the globe.

It may seem a quaint image in the age of Darwinist anthropology. As Darwin put it in *The Descent of Man*:

The main conclusion arrived at in this work [is] that man is descended from some lowly organized form. . . . There can hardly be a doubt that we are descended from barbarians. . . . Man still bears in his bodily frame the indelible stamp of his lowly origin.

One has but to note the thrust of recent works such as Jacob Bronowski's *The Ascent of Man*, Richard Leakey's *Origins* or Carl Sagan's *The Dragons of Eden* to see that with slight variations Darwinist anthropology is still the dominant model in the modern world explaining the appearance and development of man. The ancient image of a celestial origin directly challenges the assumptions of the Darwinist model and stands in the strongest possible contrast to it.

## Modern Misconceptions of Man's Origins

For some decades we have been encouraged to believe that 5000 or 6000 years ago man was everywhere a crude primitive; that he was a completely earth-born and earth-bound creature slowly evolved from an animal form. Many anthropologists and commentators have suggested that in ascending from this animal form early man was extremely vicious and aggressive. South African anatomist Raymond Dart and anthropologist L.S.B. Leakey postulate that murder was his stock in trade. As Sagan tersely puts it: 'Civilization develops not from Abel, but from Cain the murderer.'[2] In line with these speculations, contemporary cinema encourages us to believe that the first men were large, black-furred, simian-like beasts, dextrous enough to wield a stone or bone, screaming, smashing and murdering their way from one waterhole to the next.[3] The question of our origin is of supreme importance: it is the basis for our identity and destiny. The models with which we identify profoundly influence our behaviour: the man who believes he came from a beast may be more inclined to behave like a beast. The image is not only degrading; it is dangerous.

The current state of anthropology, as in so many contemporary disciplines, is perfectly summarized by the dictum of Goethe's Mephistopheles: in studying the parts, the spirit has been lost. The contemporary compartmentalization of knowledge and inquiry has produced a situation where the basic questions of anthropology remain magnificently isolated from the very developments in other fields most relevant to the nature and origin of man. This isolation has remained intact despite the 'interdisciplinary' approach of writers like Bronowski and Sagan, who do not get any further than applying findings in physical biology and computer science. They rightly see man's intelligence as the key to his evolution, but for them the intellect is locked in the tissues of the perishable brain. The focus, of

*Spiral galaxy NGC 4565, seen on edge.*

course, remains entirely on physical man, on his genes, nerves and bones. But if man is something more than merely a material being, if his intelligence and consciousness continue to exist outside of his physical body, as current research indicates, then we can hardly expect an explanation derived from purely physical remains to be comprehensive.

The specialist may object that investigations into remote viewing, out-of-body flight, pyramids, shamanism, initiations and ancient texts are out of his field. Yet most of the sources we have examined are non-technical and available in English and in other modern languages. In failing to note the relevance of such sources, the specialist merely disqualifies himself from having a complete view of the subject.

## The Spiritual Nature of Man as the 'Missing Link' in Evolution

It is indeed the isolation of most academic departments that results in the still widely held view that man was everywhere a barbarian until 6000 years ago. Ample evidence of high prehistoric civilizations now exists.[4] But while this indicates that men with high attainments have been on earth far longer than generally suspected, it does not directly bear on the question of man's origin. Ironically, much of the evidence of physical anthropology lends itself to the interpretation that man came into the earth from outside it. Skulls and fossil remains show that the emergence of physical man is unaccountably rapid by the scale of other species, and that there is not a smooth continuum from beast to man.

The very bones themselves invite the interpretation that about 200,000 years ago there was a sudden break in the interminably slow evolutionary process. In accounting for man's sudden appearance on earth, the 'missing link' is truly the spiritual nature of man. It appears that either something took over and guided the development of a preexisting animal form into man, or that through some process man simply materialized, drawing the substances of the earth into his immaterial body. In the first case, the guiding force altering the evolutionary flow can be seen as the preexisting spirits of men seeking to fashion forms suitable for them to inhabit from what was at hand. In the second, the conscious, sentient spirits may have intuited something like a Platonic form which they universally approximated, moulding their 'desire bodies' into the proper shape. This simultaneous descent and manifestation eliminates the problem of attempting to account for the development of one race from another, since in this view they came into being together, merely reflecting the varied environments of the earth. This explanation is, of course, quite magical, but in no case is the origin of man on earth likely to be less magical than out-of-body flight itself.

The testimony of the bones and stones can be argued both ways, or perhaps many ways, and can only take us so far. We have seen how man's celestial origin is testified to by ancient texts, by our contemporary behaviour, and perhaps even by the light within. It may be noted that the number of great teachers of antiquity who declared man's celestial origin is considerable. The idea, in fact, was very much alive within the Western classical tradition. Beyond these sources, we have the testimony of Eliade who found that myths relating to flights between the worlds and the celestial origin of man were literally of universal distribution. For those familiar with that scholar's enormous documentation, that—coupled with modern evidence for the actuality of out-of-body flight—is weighty testimony.

Yet there is still other evidence, evidence with implications that challenge our deepest assumptions and beliefs. Following the paths of the star maps leads us to a truly global view of human behaviour, and in many ways the price of enlightenment in these matters seems to be that we find something other than what we expected.

There is a single, all-pervading difference between the cosmologies of ancient and archaic peoples and our own: whatever may be the private personal beliefs of contemporary scientists, there is no God at the centre of modern scientific conceptions of the universe. Ever since the so called Age of Enlightenment, Western, rationalistic and materialistic science has been literally atheistic, that is without a God. When astronomer and mathematician Pierre de Laplace (1749–1827) demonstrated his theory of the solar system to Napoleon, Bonaparte is said to have asked him where God fitted into his system. Laplace responded by declaring: 'God is not necessary to the hypothesis.' In contrast, the old cosmologies, from the Egyptian to Black Elk, all pivot around the Great Spirit. The structure of their entire world view depended upon and flowed from the eternal Creator God at the centre.

We find that the 'Lord God of gods', as Joshua put it, the single Supreme Deity common to Judaism, Christianity and Islam, was known in Egypt long before the famous monotheistic pharaoh Iknaton, that is, from the earliest times until at least 1000 BC.[5] The Nesi-Khonsu Papyrus of that date states:

> This holy God [is] the Lord of all the gods . . . the holy soul who came into being in the beginning, the Great God who lives by truth, the first divine matter which gave birth unto subsequent divine matter! [He is] the being through whom every [other] god hath existence; the One One who hath made everything which has

come into existence since primeval times when the world was created; the being whose births are hidden, whose evolutions are manifold, and whose growths are unknown; the holy Form, beloved, terrible and mighty . . . the Lord of wealth, the power . . . who created every evolution of his own existence, except whom at the beginning none other existed; who at the dawn in the primeval time was Atennu, the prince of rays and beams of light . . . whose substitute is the Divine Disc.[6]

The Egyptians, then, regarded the sun as merely the substitute and symbol of its creator, the being whose births are hidden, whose growths are unknown.

It would seem that almost all religions and cosmologies agree that this Supreme God is ineffable and indescribable and that the universe is hierarchically ordered. The Supreme Deity is manifested through a series of personae, emanations, attributes, powers, avatars, lesser gods and spirits —and by a special mediator. As Taoism has it, the Tao or the Way that can be spoken of is not the Absolute Tao. The gulf between heaven and earth must be bridged by intermediaries, otherwise there could be no connection between them. In Hinduism, the Reality, the Total Godhead, is called Brahman, and as in Taoism it cannot be defined or expressed. Hinduism accepts Krishna, Buddha and Jesus as examples of divine mediators.

In the Cabbala of the Hebrews, the Ain Soph, the Ultimate is manifested in a series of emanations called the Sephiroth, the first two of which are Hockmah and Binah — Wisdom and Understanding. Hockmah or Wisdom is to the Godhead very much as Christ is to the Father, as Mithra is to Ahura-Mazda, and as Thoth is to the Supreme Deity in Egyptian cosmology. In the Book of Proverbs the Voice of Wisdom is personified and speaks of her extreme ancestry—and of man's:

The Lord created me at the beginning of his work, the first of his acts of old. Ages ago I was set up, at the first, before the beginning of the earth. . . . Before the mountains had been shaped, before the hills, I was brought forth; before he had made the earth with its fields, or the first of the dust of the world. When he established the heavens, I was there, when he drew a circle on the face of the deep . . . when he marked out the foundations of the earth, then I was beside him like a master workman; and I was daily his delight, rejoicing before him always, rejoicing in his inhabited world and delighting in the sons of men.[7]

We find that the thoroughly Biblical concept of a Supreme Father God and a Divine Son or agent through whom the world was made manifest was found among aboriginal peoples in the Americas and islands of the Pacific. In the legends of the Hopi Indians of the American Southwest, Taiowa, the Infinite Deity, first created Sotuknang, his agent and 'nephew' in order to make the world manifest. Even in the remote Pacific, the aboriginal genealogy of the Hawaiian gods begins with the relationship between Teave, the

Father-Mother Regent of the kings of heaven, and his son, Tane, which exactly parallels Christ's relationship with the Father. In the Gospel of John it is put this way:

In the beginning was the Word, and the Word was with God, and the Word was God. He was in the beginning with God; all things were made through him, and without him was not anything made that was made.[8]

In Hawaii it was taught that:

All that emanated from the Father flowed hither via his Son.[9]

## The Echo of All Religions Across Time and Space

We discover familiar images in improbable places. As he grew older, Black Elk learned he could leave his body by performing the Ghost Dance and other rituals. In a later journey he flew again into a cosmic world of Indian images and saw this:

And as I touched the ground, twelve men were coming towards me, and they said: "Our Father, the two-legged chief, you shall see!"

Then they led me to the centre of the circle where once more I saw the holy tree all full of leaves and blooming.

But that was not all I saw. Against the tree there was a man standing with arms held wide in front of him. I looked hard at him, and I could not tell what people he came from. He was not a Wasichu [white man] and he was not an Indian. His hair was long and hanging loose and on the left side of his head he wore an eagle feather. His body was strong and good to see, and it was painted red. I tried to recognize him, but I could not make him out. He was a very fine-looking man. While I was staring hard at him, his body began to change and became very beautiful with all colours of light, and around him there was light. He spoke like singing: "My life is such that all earthly beings and growing things belong to me. Your father, the Great Spirit, has said this. You too must say this."

Then he went out like a light in a wind.[10]

It has long been a subject of speculation how it is that we find such things as a cross-bearing god in Mexico, or a third century AD amulet showing a crucified god labelled Orpheus Bacchus when examples of the crucifixion in Christian art are unknown before the fifth or sixth century.[11] Many scholars have noted that the career of Jesus falls into a pattern set by Tammuz and the other so called 'vegetation gods' of the Middle East, where the death and resurrection of the god take place in the spring announcing and symbolizing the rebirth of nature. The depth of these parallels was also apparent to the peoples of antiquity:

Kings and priests from time to time made attempts to absorb the cult of Osiris into religious systems of a solar character but they failed, and Osiris, the man-god,

*A 19th century photograph of the Plains Indian ghost dance. The purpose of the dancers was to bring the two worlds closer together, and allow the dancer to cross over to the otherworld. Black Elk reports the following prayer before a ghost dance: 'Father, Great Spirit, behold these people! They shall go forth today to see their relatives, and yonder they shall be happy, day after day, and their happiness will not end.'* (Black Elk Speaks).

*Prayer before the ghost dance. (19th century photograph.)*

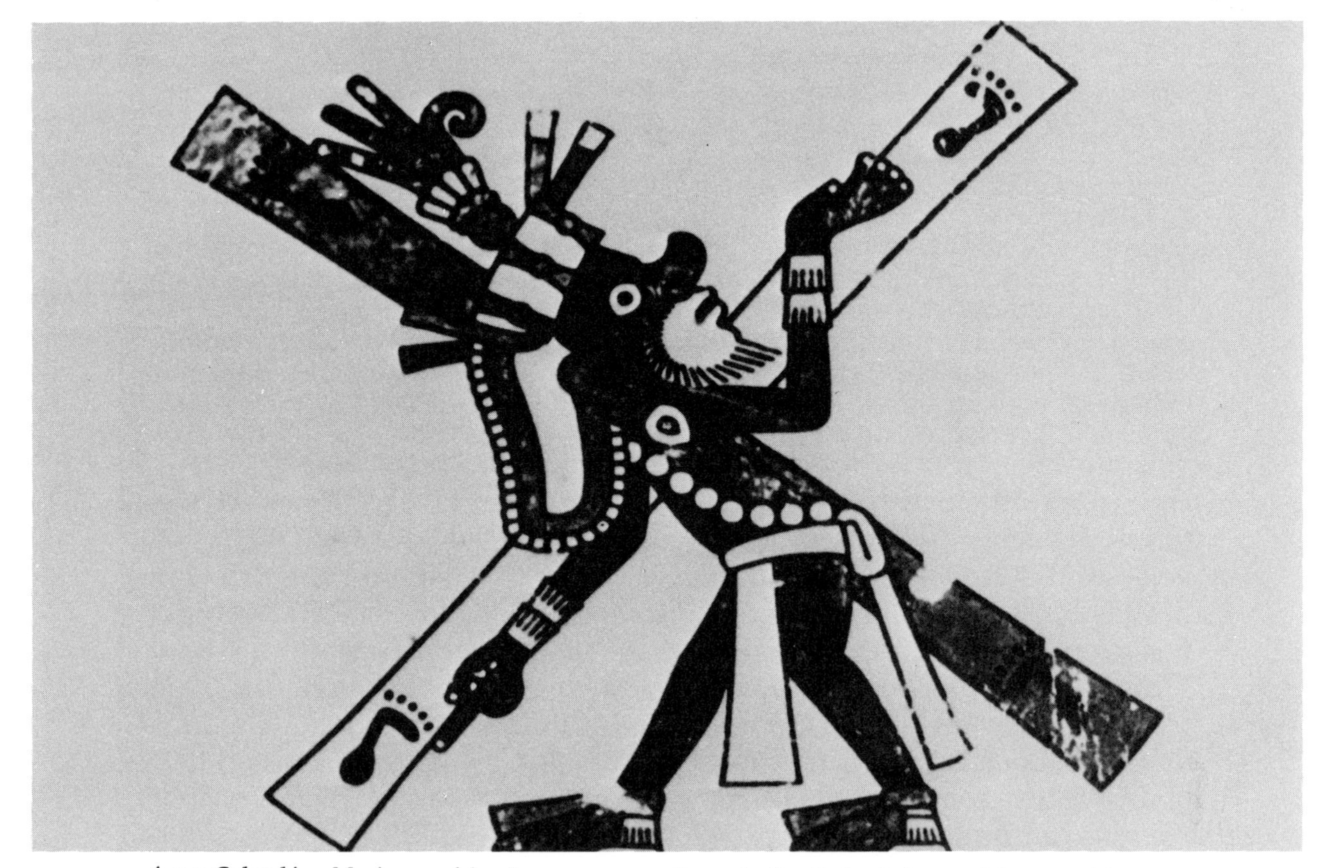

*A pre-Columbian Mexican god bearing a cross, as shown in the Codex Fejervary-Mayer, and known as Yiacatecuhtli, Lord of the Vanguard. According to Nicholson, he was in particular a god of the Pochtecas, the travelling merchants of ancient Mexico, who were followers of Quetzalcoatl.*

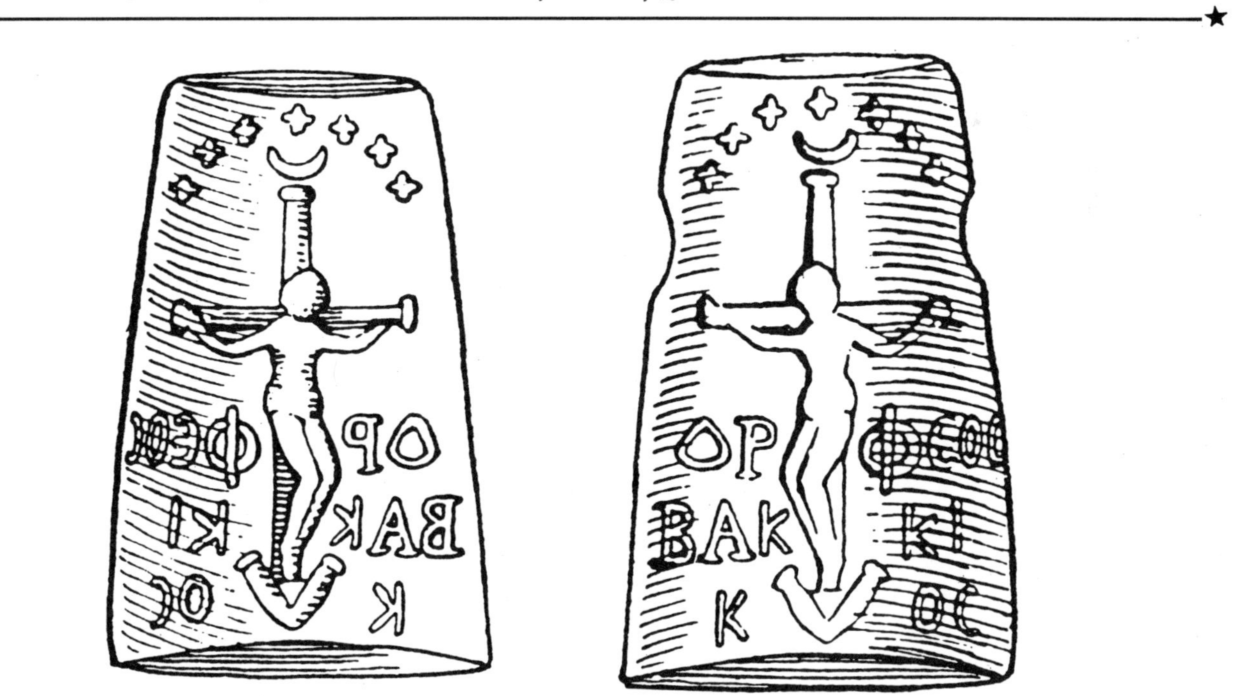


*A third century A.D. hematite seal-amulet. The impression of the seal shows a crucified figure labelled Orpheus, Bacchus. The cross is crowned with a moon and seven stars.*

always triumphed and at the last, when his cult disappeared before the religion of the Man Christ, the Egyptians who embraced Christianity found that the moral system of the old cult and that of the new religion were so similar, and the promises of resurrection and immortality in each so much alike, that they transferred their allegiance from Osiris to Jesus of Nazareth without difficulty. Moreover, Isis and the child Horus were straightway identified with Mary the Virgin and her Son.[12]

Not only was the career of Osiris, which Unas emulated, a perfect prototype of Jesus', the Mithraic, Eleusinian and related Mystery Religions all had a central hub of beliefs and practices which were forerunners of what have become Christian sacraments, themes and holy days. The Mithraics had communion, baptism and believed in a resurrection. Like Christ, Mithras was a shepherd king and a god of light. Mithraic initiations took place in the spring and 25 December was Mithras' birthday.[13] An important aspect of many ancient gods was that, like Jesus, they gave up their body. Osiris, Dionysus and Orpheus were torn apart or dismembered. Among Teutonic gods, Odin impaled himself on the World Tree and was subsequently rejuvenated. It seems to be a general rule that he who saves his life loses it, and he who loses his life, saves it. All the ancient initiations rehearse death, and the typical vision among Siberian and many other shamans during their initiatory crises was to see themselves mutilated or dismembered and then reconstituted and rejuvenated. In some cultures, initiation went beyond vision and included physical mutilation.

Studying the initiations of the Middle Eastern mystery cults one can not help but notice the profound similarities in these and, in fact, all initiations. Looking at the career of Jesus, it begins to seem a conscious and deliberate fulfillment of the role of the triumphal initiate expressed in such a way as to be comprehensible to the various initiatory cults and religions then in existence around the world. Following the pattern of torture and mutilation, Jesus lay three days in a tomb, descended into the Underworld and later ascended to heaven. The distinguishing difference between Christ and all the other initiates who made the same journey was that, instead of merely revivifying a body left in a death-like state, he resurrected a dead body and gained complete power over it, manifesting it at will.

And then we find that there is another strange pattern of convergences. Although the Creator God is ever described as unknown and hidden, the primary manifestations of this being are universally equated with light. 'God is light and in him there is no darkness at all', according to the First Letter of John. The perception of that light is the very practice of meditation. 'Go after thy sun!' Unas was told. 'Recognize the Fundamental Clear

Light.' 'Go by the Great Straight Upward Path', *The Tibetan Book of the Dead* counsels. 'I am the light of the world', Jesus said. 'Unas appears as a star'; 'a star shall come forth out of Jacob' wrote the prophet of Numbers. The beneficent angels, powers and spiritual guides reported from the days of the Old Testament prophets down to the present appear almost invariably as beings of light.

In the end we find a universal pattern linked by initiation, light and the stars. Confronted with such things as Isiac or Mithraic baptism and communion, the ill-informed Christian evangelists could only conclude that they were the work of the devil, and behaved accordingly, closing the sanctuaries. Many centuries later the veneration of a white virgin mother and of the cross as a sacred symbol among Mexican peoples practising human sacrifice so outraged the Spanish conquerors that they countered the 'sacrilege' with the tools of the Inquisition. Despite Christianity's claims to universal significance, the Church itself long obscured the truly astonishing extent to which Jesus' career built upon and fit into universal patterns of belief and practice.

Most early scholars who examined initiation had an extremely difficult time accounting for the significance of the shaman's flight, because they could not imagine that he actually experienced what he claimed. For the same reason, in analyzing the Pyramid Texts, scholars have gone to great trouble to arrive at the most accurate possible translation; then in interpreting it they disregard the very meaning arrived at. The ritual death, out-of-body flight and resurrection have usually been interpreted as so many priestly charades and tricks designed to produce dramatic appearances and amaze the credulous. Not until contemporary laboratory experiments have we been able to see that there are ample reasons to take the old initiates seriously.

Similarly, the incredible cross-cultural correspondences noted, and the way Christ fits into them, can probably be interpreted in as many ways as there are doctors of theology. But here again, perhaps the explanation is more straightforward than anything imagined. If Christ is seen as the leader sent to help man extricate himself from the sorrowful weary circle, perhaps his light has been working on the earth to achieve that end long before the days of Bethlehem.

Whether it was the work of one or of many, however it came about, the directions of the star maps, the worldwide pattern of initiation, the universal consensus on the Creator God, the perennial injunction to follow the light, the striking parallels in archaic cosmologies on opposite sides of the earth—all of this looks like the fragments of a once universal pattern. The

*Ancient depiction of Orpheus as a shepherd.*

*The death and dismemberment of Orpheus at the hands of Thracian women, as shown on a red-figured Greek vase in the Boston Museum.*

*The image of the shepherd Orpheus calming all creatures with his lyre was easily appropriated by early Christians. This portrayal of Orpheus, surrounded with biblical scenes, is from the Christian catacombs.*

*Nebula and Cluster M16.*

concept of a celestial leader sent to retrieve man from the earth ultimately only makes sense if man had come into the earth from another place, if the earth was not his home. It is just the kind of pattern which might have sprung from the simultaneous descent of millions of celestial beings into all the lands of the earth, and it hangs upon the testimony of the wise. Let those who believe that man sprang from the beasts invent their own explanations.

What kind of being is man? In the world we have made of wars and starvation, disease and pain, the words of the psalmist seem only poetry:

> They know not, neither will they understand; they walk on in the darkness; all the foundations of the earth are out of course. I have said, Ye are gods; and all of you are children of the Most High. But ye shall die like men, and fall like one of the princes.[14]

Cicero, who openly declared himself an initiate, also knew:

> It is not your outward form which constitutes your being, but your mind; not that substance which is palpable to the senses, but your spiritual nature. *Know, then, that you are a god*—for a god it must be which flourishes, and feels, and recollects, and foresees, and governs, regulates and moves the body over which it is set, as the Supreme Ruler does the world which is subject to him.[15]

Everything man does on earth necessarily proceeds from the metaphysical to the physical, from the world of thought to the world of matter and action. Anything and everything man does, he does first in thought, a natural consequence of the fact that man is primarily a mental-spiritual being. Thought affects and moulds and changes matter. Thought has dominion over matter. It is by virtue of his mental-spiritual character that man rules the earth, building the patterns that manifest in crimes or miracles and in our habitations in the world beyond. And though it seems a fate beyond our dreaming, when we have overcome the earth, the destiny of star-born man is among the lights of heaven, in Edens in the stars, where we may be the gods and goddesses that govern, lead and light the way in the primeval gardens of beginnings.

> And while all other creatures from their birth
> With downcast eyes gaze on their kindred earth;
> He bids man walk erect, and scan the heaven
> From whence his soul has sprung, to which his hopes are given.[16]

*The relief on the 3.65 metre (12 foot) limestone sarcophagus lid found in 1952 under the pyramid-shaped Temple of Inscriptions at Palenque, Mexico. It is thought to represent Pacal, a seventh century ruler at Palenque. Although it has been subject to many interpretations, the central image is clearly that of a cross with a two-headed serpent entwined around its arms; at the top it is surmounted by a bird.*

CHAPTER ONE
**The Stratagems of Time**

1. Caesar, *De Bello Gallico*, Book VI, Chapter XVI.
2. G.S. Hawkins, *Stonehenge Decoded, Beyond Stonehenge.*
3. Alexander Thom, *Megalithic Sites in Britain, Megalithic Lunar Observatories.*
4. John Eddy, 'Mystery of the Medicine Wheels', *National Geographic,* December, 1977; E.C. Krupp, *In Search of Ancient Astronomies*; and Mark Feldman, *The Mystery Hill Story.*
5. L.C. Stecchini, 'Notes on the Relation of Ancient Units of Measure to the Great Pyramid', an appendix in Peter Tompkins, *Secrets of the Great Pyramid.*

CHAPTER TWO
**A 4000 Year Old Picture**

1. De Santillana and von Dechend, *Hamlet's Mill,* p. 119.
2. J.H. Breasted makes this point several times in *The Dawn of Conscience.*
3. Among other powers, Set was also thought to be the cause of clouds, mist, rain, thunder and lightning, hurricanes and storms, earthquakes and eclipses.
4. H.B. Fell, *America B.C.*, pp. 261-268.
5. Hieroglyphic texts contain no vowels and since not all consonants are represented in parallel texts in other languages, such as those on the Rosetta stone, the phonetic equivalents of many words can only be guessed at. As in many other cases, Egyptologists do not agree on a single spelling or pronunciation for 'Djed'. Zed, Zet, Tet, Sed and similar variations are found from writer to writer for the same word.
6. *Hamlet's Mill,* p. 232.
7. Wm. H. Stahl, *Macrobius' Commentary on the Dream of Scipio,* Appendix A, p. 249-250.
8. Wallis Budge, *The Book of the Dead,* Vol. III, p. 647.

CHAPTER THREE
**Star Temples**

1. In times long after the building of the pyramids, around 600 B.C., it became a fad to use some pyramids for burials. These later internments, especially numerous in the Step Pyramid at Saqqara, are referred to as 'intrusive burials'.
2. Aside from the sarcophagus itself, which is nicely finished, the masonry of this pyramid is exceptionally crude. The excavating archaeologist decoded the name Sekhemket from jar sealings found in the substructure. Although the excavator, Zakaria Ghoneim, attempted to identify this name with a short-lived pharaoh of the First or Second Dynasty, there is little to indicate who or what Sekhemket was. Nor is it clear whether the pyramid is among the earliest or the latest constructed.
3. See *Pyramid Odyssey*, Chapter 7, for a fuller discussion of the tomb theory.
4. The Great Pyramid is located at 29 degrees 58 minutes 51 seconds (29° 58′ 51″) north latitude. In order to point directly at the celestial pole, the passage would have to describe this same angle from ground level at Gizeh. Various reports of the angle of the Descending Passage range from 26° 2′ 30″ to 26° 34′, a difference of some three degrees. Actually, this is a descending passage only from the point of view of a person *entering* the Pyramid. For someone leaving the Pyramid by the same route, it would be an *ascending* passage.

5. See *Pyramid Odyssey*, Chapters 8 and 11.
6. Norman Lockyer, *The Dawn of Astronomy*, p. 302.
7. This identification arises from the fact that the Circumpolar Stars typify the night sky because they do not rise or set but are always above the horizon. Lockyer also noted that Horus seemed to be identified with almost any planet or constellation (never a single star) *at the moment of rising*. These observations concerning Set and Horus further confirm de Santillana and von Dechend's interpretation of Senusert's monument in which these gods turn the churn of heaven.
8. Lockyer, *The Dawn of Astronomy*, p. 310, p. 354. Since the slow wobble of the earth's axis that produces the shift of the celestial pole (the pole of the equator) is a circle with a radius of approximately 23.5 degrees from the stationary pole of the ecliptic, a star once the Pole Star 13,000 years later may be as much as 47 degrees (23.5 x 2 = 47) distant from the pole. At 30 degrees north latitude — the approximate latitude of Gizeh — the Pole Star is only 30 degrees above the horizon. Thus, some stars, like Eltanin, may at one time be the Pole Star and thousands of years later will rise from the horizon in the northeast.
9. Lockyer, *The Dawn of Astronomy*, p. 100.
10. Ibid, p. 108.
11. Ibid.
12. Ibid, p. 177.
13. Ibid, p. 176.
14. Ibid, pp. 352-354.
15. Ibid, p. 146.

## CHAPTER FOUR
### Star Maps

1. Wallis Budge, *The Gods of the Egyptians*, Vol. I, pp. 170-179.
2. Wallis Budge, *Book of the Dead*, Vol. I, pp. xxxii-xxxiii.
3. Ibid, Vol. II, p. 320.
4. Ibid, Vol. II, pp. 440-443; and Budge, *The Gods of the Egyptians*, Vol. I, pp. 176-177.
5. O. Neugebauer and Richard A. Parker, *Egyptian Astronomical Texts*, Vol. III, *Decans, Planets, Constellations and Zodiacs*.
6. At the time these star clocks were designed, the Egyptians divided both the night and day into 12 hours each, even when the night and day were of different lengths. The star clocks, therefore, show which star or stars were rising for each 12 divisions of the night, and they record this information for 10 day intervals (the old Egyptian week) throughout the year. This interval imposed (or reflected) a patterning of the sky into 10 degree sections, called decans, each decan identified by a certain star or group of stars. Since there are 36.5 10 day intervals and 12 divisions of the night, they used a basic grid 37 columns long and 12 columns high to record the shifting rising times of the decans, which consequently form distinguishing diagonal patterns. While the basic pattern was 37 by 12, there are often extra columns to record ancillary information, or abbreviated versions using star groupings covering more than 10 degrees. See Neugebauer and Parker, *Egyptian Astronomical Texts*, Vol. I, *The Early Decans*; also E.C. Krupp, *In Search of Ancient Astronomies*, pp. 208-209.
7. In later times, during the era of Ramses in the 13th century BC, another form of star

clock was developed in which stars were recorded as they culminated, or passed the meridian, an imaginary north-south line directly overhead. Again, the basic interval used in these observations was 10 days, since at a particular hour a particular star would be in about the same position for that period of time. Here, however, the sightings were recorded on a grid relative to the body of an assistant who sat in front of the observer. These Ramesside star clocks are not found inside coffins but on the walls of tombs.

8. Not all references and symbols in the star maps are understood. In some cases, the partial condition of the monuments, in others, the highly abbreviated form of the information prevents full decipherment. The maps may employ temporal synchronization between certain elements which has escaped us. Only since the publication of Volume III of Neugebauer and Parker's *Egyptian Astronomical Texts* in 1969 have many of these star maps been easily available even to scholars. Many of the monuments and remains recording astronomic information are in a partial or ruinous condition, and have deteriorated since they were recorded. Some have disappeared. Those that are extant are scattered in museums around the world. The best and in some cases the only records of monuments are now found in scholarly works, of which *Egyptian Astronomical Texts* is by far the most complete and important collection.

## CHAPTER FIVE
### The Star Walker

1. Utterance 284, spell 424, Alexandre Piankoff, *The Pyramid of Unas*, p. 49.
2. Utterance 228, spell 288, Piankoff, p. 95.
3. Utterance 235, spell 239, Piankoff, p. 97. Actually, Piankoff's full version of this utterance is: 'Thou! Ah, ah, ah! Filler! Filler! Thou art to rape the two [holes] of the [door] stone, door jamb! The two which are! I, ia, i!'
4. Piankoff, p. 97, note on utterance 235, cites Spiegel, 'Arbeiterrenden in den Pyramidentexten,' *Orientalia*, (Rome), XXII, 1953, pp. 233-241.
5. Utterance 317, spells 508-510, Piankoff, p. 18.
6. Utterance 219, spell 167, Piankoff, p. 64.
7. Utterance 219, spells 192-193, Piankoff, p. 68.
8. Utterance 220, spell 194, Piankoff, p. 68.
9. Utterance 222, spell 201, Piankoff, p. 70.
10. Utterance 222, spell 202, Piankoff, p. 70.
11. Utterance 222, spell 207, Piankoff, p. 70.
12. Utterance 222, spell 208, Piankoff, p. 70.
13. Utterance 222, spells 209-210, Piankoff, pp. 70-71.
14. Utterance 210, spell 130, Piankoff, p. 74.
15. R.O. Faulkner, 'The King and the Star Religion in the Pyramid Texts', *Journal of Near Eastern Studies*, p. 154.
16. Faulkner, ibid.
17. Utterance 269, spell 380, Piankoff, p. 42. Also see Faulkner in the work cited. Piankoff, in fact, uses 'imperishable stars' for 'circumpolar stars'. Faulkner is more specific, translating 'circumpolar stars' in the same place. The circumpolar stars are 'imperishable' since they do not rise nor set.
18. Utterance 214, spell 138, Piankoff, p. 60. Same comment as in previous note.

19. Spells 138-139, Faulkner, work cited, p. 155.
20. Utterance 222, spell 212, Piankoff, p. 71.
21. Utterance 213, spell 134, Piankoff, p. 59.
22. Utterance 215, spell 145, Piankoff, p. 60.
23. J.H. Breasted, *Development of Religion and Thought in Ancient Egypt*, p. 39.
24. A. Moret, *Mystères Egyptiens*, pp. 187–190, as quoted by Sir James G. Fraser, *Adonis Attis Osiris*, third edition, p. 155.
25. Flinders Petrie, *A History of Egypt*, (From the Earliest Kings to the XVIth Dynasty), p. 144. Petrie commented: 'All that we know of this king (Neb-tau-ra or Mentuhetep IV) is from the inscriptions cut by the working parties in the quarries of Hammamat. We find that in his second year a Sed festival of Sirius' rising took place, another instance which shows that these festivals were then at fixed astronomic dates, and not dependent on the year of the reign. Most of these inscriptions relate to the party who prepared the royal sarcophagus . . . in the second year of the king's reign.'
26. S. Hassan, *Excavations at Giza*, Vol. 6, Part I, p. 66.
27. Utterance 275, spell 416, Piankoff, p. 47.
28. Utterance 223, spell 214, Piankoff, p. 71.
29. Utterance 223, spell 216, Piankoff, p. 71.
30. Utterance 223, spell 217, Piankoff, p. 71.
31. Utterance 224, spell 218, Piankoff, p. 72.
32. Utterance 224, spell 221, Piankoff, p. 72.
33. Utterance 223, spell 214, Piankoff, p. 71.
34. Utterance 127, spell 80, Piankoff, p. 89.
35. Utterance 223, spell 214, Piankoff, p. 71.
36. Egyptological opinion is divided on whether it is the *ka* or the *ba* that remains behind. The New Kingdom Egyptians used these terms so interchangeably that they themselves may have lost the original distinction between them.
37. Utterance 154, spell 92, Piankoff, p. 92.
38. Utterance 260, spell 316, Piankoff, p. 36.
39. Utterance 264, spell 350. See Selim Hassan, *Excavations at Giza*, Vol. VI, Part I, p. 29.
40. See, for example, Elisabeth Haich, *Initiation*.

## CHAPTER SIX
### On the Trail of Unas

1. Russell Targ and Harold Putoff, *Mind-Reach*, pp. 208-209.
2. Ibid, p. 211.
3. Ibid, p. 169, cites editorial, 'Scanning the Issue', in *Proceedings of the IEEE*, LXIV, (March 1970) No. 3.
4. C.G. Jung, *Memories, Dreams, Reflections*, pp. 289-290.
5. For the connection between quantum physics and parapsychology, see 'Quantum Physics and Parapsychology', *Parapsychology Review*, Vol. 5, No. 6, November/ December 1974.
6. In his own work, *To Kiss Earth Good-bye*, Swann seems more inclined to see it as out-of-body experience.
7. Mircea Eliade, *Myths, Dreams and Mysteries*, p. 87.
8. Charles T. Tart, *PSI*, pp. 196-198.

9. Targ and Putoff, pp. 191-193, cite *Proceedings of the Society for Psychical Research* (London, 1891-1892), Vol. VII, p. 41.
10. Norma Bowles and Fran Hynds with Joan Maxwell, *Psi SEARCH*, pp. 62-63.

## CHAPTER SEVEN
### Initiation

1. Eliade, *Shamanism*, p. 4.
2. Ibid, p. 5.
3. Ibid, p. 265.
4. Ibid, p. 259.
5. Ibid, p. 260.
6. Ibid, p. 261.
7. Ibid, p. 259.
8. Joseph L. Henderson and Maud Oakes, *The Wisdom of the Serpent*, p. 206.
9. Eliade, *Shamanism*, p. 264.
10. Plutarch, *Of Isis and Osiris*, pp. 27 and 35.
11. Nilsson, *Greek Folk Religion*, p. 59, cites Pindar, Fragment 137, in T. Bergk, *Poetae lyrici Graeci*, fourth edition, Leipzig, 1878-1882.
12. Eliade, *Patterns in Comparative Religion*, p. 426.
13. *The Golden Ass of Apuleius*, translated by Robert Graves, p. 252.
14. Eliade, *Myths, Dreams and Mysteries*, p. 110.
15. J.R. Porter, 'Muhammad's Journey to Heaven', *The Journey to the Other World*, edited by H.R. Ellis Davidson, p. 16.
16. As per Eliade, *Rites and Symbols of Initiation*, p. 111.
17. G. Mylanos, *Eleusis and the Eleusinian Mysteries*, p. 264.
18. *The Golden Ass*, Book XI.
19. Eliade, *Myths, Dreams and Mysteries*, p. 205.
20. I, Nicholson, *Mexican and Central American Mythology*, p. 88.
21. Richard F. Dempewolff, 'Putting Biblical Pieces Together', *Popular Mechanics*, May 1978, p. 103 ff.
22. Nicholson, work cited, pp. 68-70.
23. S. Grof and J. Halifax, *The Human Encounter with Death*, p. 183.
24. Eliade, *Shamanism*, p. 417.
25. *Larousse Encyclopaedia of Mythology*, 1960 edition, p. 313.
26. *Newsweek*, 1 May 1978, p. 53.
27. Cicero, *De Legibus*, II.
28. Alexandre Moret, *Kings and Gods of Egypt*, pp. 194-195.
29. *Black Elk Speaks*, pp. 208-209.
30. Utterance 204, spell 118, Piankoff, p. 72.
31. James H. Breasted, *Ancient Records of Egypt*, Vol. III, pp. 230-231, art. 549-551.
32. As per John Michell, *City of Revelation*, p. 16.

## CHAPTER EIGHT
### The Star Ships

1. G.R.S. Mead, *Thrice Greatest Hermes*, III, p. 30.
2. See note 34, Chapter 5.
3. See especially, M.F. Long, *The Secret Science Behind Miracles*, pp. 137-140.

4. W.Y. Evans-Wentz, *The Tibetan Book of the Dead*, p. 158.
5. Ibid.
6. Ibid, p. 158, note 3.
7. Raymond A. Moody, *Life After Life*, p. 45.
8. Raymond A. Moody, *Reflections on Life After Life*, p. 16.
9. George G. Ritchie, *Return from Tomorrow*, pp. 38-39.
10. Moody, *Life After Life*, pp. 41-42.
11. Stanislav Grof and Joan Halifax, *The Human Encounter with Death*, p. 159.
12. S.G.F. Brandon, *The Judgment of the Dead*, p. 175.
13. See Moody, *Reflections on Life After Life*, pp. 32-33; also see Ritchie, *Return from Tomorrow*, p. 54.

## CHAPTER NINE
### The Rivers of Heaven

1. Raymond A. Moody, *Reflections on Life After Life*, pp. 15-16.
2. Ibid, p. 16.
3. Ibid, pp. 16-17.
4. Ibid, p. 74.
5. *The Divine Comedy*, Inferno, Canto 3, White translation.
6. *The Republic of Plato*, F.M. Cornford translation, pp. 348-359.
7. De Santillana and von Dechend, *Hamlet's Mill*, p. 188.
8. *Theogony*, 790 ff.
9. See W.Y. Evans-Wentz, *The Tibetan Book of the Dead*, p. 62.
10. *Georgics*, I, 242 f.
11. *Hamlet's Mill*, pp. 193-194.
12. Ibid, pp. 192-193.
13. Ibid, p. 188.
14. Ibid, p. 194.
15. Ibid, p. 196.
16. Ibid, p. 188.
17. Ibid, p. 196.
18. *Brihad-Aranyaka Upanishad*, 5. 10, *The Thirteen Principal Upanishads*, translated by R.E. Hume, p. 139.
19. Plutarch, *Selected Lives and Essays*, 'On Exile', translated by Louise Ropes Lomis, pp. 364-365.
20. *Aeneid*, Book VI.
21. See, for example, Matthew 11:12-15; 16:13-14; 17:10-13.
22. See Ian Stevenson, *Twenty Cases Suggestive of Reincarnation*.
23. Lucan, *Pharsalia*, 1. 458-462; as per W.Y. Evans-Wentz, *The Fairy Faith in Celtic Countries*, p. 368.
24. As reported by Joseph L. Henderson and Maud Oakes, *The Wisdom of the Serpent*, pp. 209-210.

## CHAPTER TEN
### The Golden Plate of Petelia

1. *The Republic*, Book VII, 514-521.
2. *Black Elk Speaks*, p. 85.
3. W.K.C. Guthrie, *Orpheus and Greek Religion*, pp. 171-182.

4. Ibid.
5. *Paradisio,* Canto 14, Lawrence G. White translation.
6. Koran, LVI, 1-55.
7. Ibid.
8. Guthrie, work cited.
9. Ibid.
10. Cicero, *De Republica*, Book IV, XV-XVI, translation of C.D. Yonge, *Cicero's Nature of the Gods*, p. 383.
11. Eliade, *Myths, Dreams and Mysteries*, pp. 104-105.
12. The Dream of Scipio, Chapter III; see Wm. H. Stahl, *Macrobius' Commentary on the Dream of Scipio*, p. 72.
13. J. Eric S. Thompson, *Maya Hieroglyphic Writing*, p. 85.
14. See Gershom G. Scholem, *Major Trends in Jewish Mysticism*, pp. 244-286.

## CHAPTER ELEVEN
**The Lake of Memory**

1. Stanislav Grof and Joan Halifax, *The Human Encounter with Death*, p. 177.
2. *Black Elk Speaks*, p. 43.
3. John C. Lilly, *The Center of the Cyclone*, p. 212.
4. *New Age,* November 1977, pp. 82-83.
5. See Wm. James, *The Varieties of Religious Experience*, pp. 410-411.
6. Robert A. Monroe, *Journeys Out of the Body*, pp. 123-126.
7. Ritchie, *Return from Tomorrow*, p. 69.
8. Ibid.
9. *Divine Comedy*, Inferno, Canto 14, Lawrence G. White translation.
10. Ritchie, work cited, pp. 63-64.

## CHAPTER TWELVE
**The Gods of Light**

1. Job 38:19-21, RSV.
2. Carl Sagan, *The Dragons of Eden*, p. 98.
3. See Stanley Kubrick's *2001: A Space Odyssey*, opening scene.
4. See Fix, *Pyramid Odyssey*, Chapter 14.
5. Budge, *The Gods of the Egyptians*, Vol. I, pp. 137-144.
6. Budge, *Book of the Dead*, Vol. III, pp. 645-648.
7. Proverbs 8:22-31, RSV.
8. John 1:1-3, RSV.
9. Leinani Melville, *Children of the Rainbow*, p. 18.
10. *Black Elk Speaks*, p. 249.
11. W.K.C. Guthrie, *Orpheus and Greek Religion*, p. 265 ff.
12. Budge, *The Gods of the Egyptians*, Vol. I, p. xv.
13. Franz Cumont, *The Mysteries of Mithra*, p. 167.
14. 82nd Psalm, 5-7, KJV.
15. Cicero, *On the Republic*, Book VI, XXIV.
16. Ovid, *Metamorphoses*, I, 86.

# WORKS CONSULTED

Abell, George O. *Exploration of the Universe*. New York: Holt Rinehart & Winston, 1975.

Alighieri, Dante. *The Divine Comedy*. Translated by Lawrence Grant White. New York: Pantheon Books, 1948.

Allen, Richard Hinckley. *Star Names: Their Lore and Meaning*. New York: Dover, 1963. Reprint of 1899 edition.

Apuleius, Lucius. *The Golden Ass of Lucius Apuleius*. Translated by William Adlington. London: Murray's Book Sales Ltd. No date.

*The Golden Ass of Apuleius*. Translated by Robert Graves. New York: Farrar Strauss & Young, 1951.

The Bible, King James and Revised Standard versions.

Black Elk. See Neihardt.

Bleeker, C.J. *Initiation*. Leiden, The Netherlands: E.J. Brill, 1965.

*Hathor and Thoth*. Leiden, The Netherlands: E.J. Brill, 1973.

Bowles, Norma, and Fran Hynds. *Psi SEARCH*. New York: Harper & Row, 1978.

Brandon, S.G.F. *The Judgment of the Dead*. New York: Charles Scribner's Sons, 1975.

Breasted, James Henry. *Development of Religion and Thought in Ancient Egypt*. New York: Charles Scribner's Sons, 1912.

*The Dawn of Conscience*. New York: Charles Scribner's Sons, 1933.

*Ancient Records of Egypt*, five volumes. New York: Russell & Russell, Inc., 1962.

Bronowski, Jacob. *The Ascent of Man*. Boston and Toronto: Little Brown and Company, 1973.

Budge, E.A. Wallis. *The Book of the Dead*, three volumes. London: Kegan Paul, Trench, Trübner & Co., Ltd., 1901.

*The Gods of the Egyptians*, two volumes. New York: Dover Publications, Inc., 1969. Reprint of 1904 edition.

*Osiris & The Egyptian Resurrection*, two volumes. New York: Dover Publications, Inc., 1973. Reprint of 1911 edition.

Caesar, Julius. *De Bello Gallico*. Translated by H.J. Edwards in *The Gallic War*. Cambridge, U.S.A.: Harvard University Press; London, England: William Heinemann Ltd., 1970.

*De Bello Gallico*. Literally translated in *Caesar's Commentaries*. London: Henry G. Bohns, 1851.

Cicero, M. Tullius. *De Republica*, or *On the Commonwealth*. *De Legibus*, or *On the Laws*. *De Naturo Deorum* or *Of the Nature of the Gods*. Translated by C.D. Yonge, in *Cicero's Nature of the Gods*. London: George Bell & Sons, 1907.

Cole, J.H. *The Determination of the Exact Size and Orientation of the Great Pyramid at Giza*. Cairo: Government Press, 1925.

Crookall, Robert. *Out-of-the-Body Experiences: A Fourth Analysis*. New York: University Books, Inc., 1970.

*Case-book of Astral Projection, 545-746*. Secaucus, U.S.A.: University Books, Inc., 1972.

Cross, Tom Peete, and Clark Harris Slover, ed. *Ancient Irish Tales*. New York: Henry Hold and Company, 1936.

Cumont, Franz. *The Mysteries of Mithra*. New York: Dover Publications, Inc., 1956.

Darwin, Charles. *The Descent of Man*, second edition. New York: Hurst & Company. No date.

Davidson, H.R. Ellis, ed. *The Journey to the Other World*. Cambridge, England: D.S. Brewer Ltd., 1975. Totowa, U.S.A.: Rowman and Littlefield, 1975.

Davis, Whitney M. 'The Ascension-Myth in the Pyramid Texts.' *Journal of Near Eastern Studies*, Vol. 36, No. 3, July 1977.

Dechend, Hertha von. See Santillana.

Dempewolff, Richard F. 'Putting Biblical Pieces Together.' *Popular Mechanics*, May 1978.

Dobyns, Henry F. See Euler.

Dods, Marcus. *Forerunners of Dante*. Edinburgh: T. & T. Clark; London: Simpkin, Marshall, Hamilton, Kent, and Co. Limited; New York: Charles Scribner's Sons, 1903.

Duran, Fr. Diego. *Book of the Gods and Rites and The Ancient Calendar*. Translated by Fernando Horcasitos and Doris Heyden. Norman: University of Oklahoma Press, 1971.

Eddy, John A. 'Probing the Mystery of the Medicine Wheels.' *National Geographic*, Vol. 151, No. 1, January 1977.

Edwards, I.E.S. *The Pyramids of Egypt*. New York: Pitman, 1961.

Eliade, Mircea. *The Myth of the Eternal Return*. Princeton: Princeton University Press, 1954.

*Rites and Symbols of Initiation*. New York: Harper & Row, 1965. (First published by Harper & Brothers, 1958.)

*Myths, Dreams and Mysteries*. New York: Harper & Row, 1967. (©1960 in English translation by Harvill Press.)

*Shamanism*. Princeton: Princeton University Press, 1972.

*Patterns in Comparative Religion*. New York: New American Library, 1974. (©1958, Sheed & Ward, Inc.)

*Death, Afterlife, and Eschatology*. New York: Harper & Row, 1974.

Erman, Adolf. *Life In Ancient Egypt*. New York: Dover Publications, Inc., 1971. (Reprint of 1894 edition.)

Euler, Robert C., and Henry F. Dobyns. *The Hopi People*. Phoenix: Indian Tribal Series, 1971.

Evans-Wentz, W.Y. *The Fairy-Faith in Celtic Countries*. New York: University Books, Inc., 1966.

*The Tibetan Book of the Dead*. London, Oxford, New York: Oxford University Press, 1978.

Faulkner, R.O. 'The King and the Star Religion in the Pyramid Texts.' *Journal of Near Eastern Studies*, Vol. XXV, No. 3, July 1966.

Feldman, Mark. *The Mystery Hill Story*. North Salem, U.S.A.: Mystery Hill Press, 1977.

Fell, Barry. *America B.C.* New York: Quadrangle/The New York Times Book Co., 1976.

Fix, Wm. R. *Pyramid Odyssey*. New York: Mayflower Books, 1978.

Frazer, James George. *Adonis Attis Osiris*, third edition. New Hyde Park, New York: University Books, 1961.

Gardiner, Alan. *Egyptian Grammar*, third edition. Oxford: Griffith Institute, Ashmolean Museum, 1976.

Gosse, A. Bothwell. *The Civilization of the Ancient Egyptians*. London: T.C. & E.C. Jack, Ltd. No date.

Greenhouse, Herbert B. *The Astral Journey*. New York: Avon Books, 1976 (by arrangement with Doubleday & Co., 1974.)

Grof, Stanislav, and Joan Halifax. *The Human Encounter With Death*. New York: E.P. Dutton, 1977.

Guirard, Felix, ed. *Larousse Encyclopedia of Mythology*. New York: Prometheus Press, 1960. London: Batchworth Press Limited, 1959.

Guthrie, W.K.C. *Orpheus and Greek Religion*, second edition. London: Methuen & Co.

Ltd., 1952.
Haich, Elizabeth. *Initiation*. London: George Allen & Unwin Ltd., 1965.
Halifax, Joan. See Grof.
Hamblin, Dora Jane, and the editors of Time-Life Books.
*The Etruscans*, The Emergence of Man series. New York: Time-Life Books, 1975.
Hapgood, Charles. *The Earth's Shifting Crust*. New York: Pantheon, 1958. Later published as *The Path of the Pole*. New York: Chilton Book Company, 1970.
*Maps of the Ancient Sea Kings*. Philadelphia: Chilton Book Company, 1966.
Haraldsson, Elendur. See Osis.
Hassan, Selim. *Excavations at Giza*, Vol. VI, pt. 1. Cairo: Government Press, 1946.
Hawkins, Gerald S. *Stonehenge Decoded*. Garden City, New York: Doubleday, 1965.
*Beyond Stonehenge*. London: Hutchinson, 1973.
Henderson, Joseph L., and Maud Oakes. *The Wisdom of the Serpent*. New York: Collier Books, 1971.
Herodotus. *The Histories*. Translated by Aubrey de Selincourt. Baltimore, U.S.A.: Harmondsworth, England; Ringwood, Victoria, Australia: Penguin Books, 1968.
Hesiod. *Theogony*. Translated by Dorothea Wender in *Hesiod and Theognis*. Baltimore, U.S.A.; Harmondsworth, England; Ringwood, Victoria, Australia: Penguin Books, 1973.
Hoyle, Fred. 'Stonehenge—An Eclipse Predictor.' *Nature*, 211, 30 July 1966.
Hull, Eleanor, ed. *The Cuchullin Saga*. London: David Nutt, 1898.
Hynds, Fran. See Bowles.
Ingrasci, Rick. See Taylor, Peggy.
James, William. *The Varieties of Religious Experience*. London and New York: Longmans, Green and Co., 1911.
Jung, C.G. *Memories, Dreams, Reflections*. New York: Random House, 1963.
Kletti, Roy. See Noyes.
Koran. Translated by J.M. Rodwell. London and Toronto: J.M. Dent & Sons Ltd. New York: E.P. Dutton Co., 1926. First published in 1909.
Krupp, E.C. *In Search of Ancient Astronomies*. Garden City, New York: Doubleday & Company, Inc., 1977.
Kübler-Ross, Elisabeth, ed. *Death: The Final Stage of Growth*. Englewood Cliffs, U.S.A.: Prentice-Hall, Inc., 1975.
Kyselka, Will, and Ray Lanterman. *North Star to Southern Cross*. Honolulu: The University Press of Hawaii, 1976.
La Farge, Oliver. *A Pictorial History of the American Indian*. New York: Crown Publishers, Inc., 1956.
Lanterman, Ray. See Kyselka.
Lauer, Jean-Philippe. *Les Pyramides de Sakkarah*, fourth edition. Cairo: L'Institut Français d'Archéologie Orientale, 1972.
*Saqqara: The Royal Cemetery of Memphis*. New York: Charles Scribner's Sons; London: Thames and Hudson, Ltd., 1976.
Lesko, Leonard H. 'Some Observations on the Composition of the Book of Two Ways.' *Journal of the American Oriental Society*, 91.1 (1971).
Lilly, John C. *The Center of the Cyclone*. New York: The Julian Press, Inc., 1973.
Lockyer, J. Norman. *The Dawn of Astronomy*. New York and London: Macmillan and Co., 1894.

Long, Max Freedom. *The Secret Science Behind Miracles*. Los Angeles: Huna Research Publications, 1954.
*The Huna Code In Religions*. Santa Monica: De Vorss & Co., 1965.
*Mabinogian, The*. Translated by Lady Charlotte Guest. London: J.M. Dent & Sons Ltd.; New York: E.P. Dutton & Co. Inc., 1937. First published in 1906.
Macrobius. *Commentary on the Dream of Scipio*. Translated with introduction and notes by William Harris Stahl. New York and London: Columbia University Press, 1952.
Mark, Rachel. See Woodward.
Marshack, Alexander. *The Roots of Civilization*. New York: McGraw Hill, 1972.
Matson, Archie. *Afterlife: Reports From the Threshold of Death*. New York: Harper & Row, 1975.
Mead, G.R.S. *Thrice Greatest Hermes*, three volumes. London: John M. Watkins, 1949.
Melville, Leinani. *Children of the Rainbow*. Wheaton, U.S.A.: Theosophical Publishing House, 1969.
Mendelssohn, Kurt. *The Riddle of the Pyramids*. New York: Praeger Publishers; London: Thames & Hudson, 1974.
Mercer, Samuel A.B. *The Pyramid Texts in Translation and Commentary*, Vols. 1 and 2. New York, Toronto, London: Longmans, Green and Co., Inc., 1952.
Michell, John. *City of Revelation*. London: Abacus, 1973.
Monroe, Robert A. *Journeys Out of the Body*. Garden City, New York: Doubleday & Company, Inc., 1971.
Moody, Raymond A. *Life After Life*. Covington, U.S.A.: Mockingbird Books, 1975.
*Reflections on Life After Life*. St. Simons Island, Georgia, U.S.A.: Mockingbird Books, 1977.
'Cities of Light.' *New Age*, November 1977.
Moret, Alexandre. *Kings and Gods of Egypt*. New York: G.P. Putnam's Sons, 1912.
Mylonas, George E. *Eleusis and the Eleusinian Mysteries*. Princeton: Princeton University Press, 1961.
Neihardt, John G. *Black Elk Speaks*. Lincoln: University of Nebraska Press, 1961.
Neugebauer, O. *The Exact Sciences in Antiquity*, second edition. New York: Harper & Brothers, 1962.
Neugebauer, O., and Richard A. Parker. *The Early Decans*, Vol. I of *Egyptian Astronomical Texts*. London: Lund Humphries, 1960. Published for Brown University Press, Providence, Rhode Island.
*Decans; Planets, Constellations and Zodiacs* (parts 1 and 2, text and plates), Vol. III of *Egyptian Astronomical Texts*. Providence, Rhode Island: Brown University Press; London: Lund Humphries, 1969.
*Newsweek*. 'Living With Dying.' 1 May 1978, Vol. XCI, No. 18.
Nicholson, Irene. *Mexican and Central American Mythology*. London, New York, Sydney, Toronto: Paul Hamlyn, 1967.
Nilsson, Martin P. *Greek Folk Religion*. The Torchbook edition. New York: Harper & Row, 1961. First published under the title *Greek Popular Religion*. New York: Columbia University Press, 1940.
Noyes, Russell, and Roy Kletti. 'The Experience of Dying from Falls.' *Omega*, Vol. 3, 1972.
Oakes, Maud. See Henderson.
Osis, Karlis. *Deathbed Observations by Physicians and Nurses*. New York: Parapsychology Foundation, Inc., 1961.

Osis, Karlis, and Elendur Haraldsson. *At the Hour of Death*. New York: Avon Books, 1977.

Palmer, John and Carol Vassar. 'ESP and Out-of-the-Body Experiences: An Exploratory Study.' *Journal of the American Society for Psychical Research*, Vol. 68, No. 3, July 1974.

Panati, Charles. 'Quantum Physics and Parapsychology.' *Parapsychology Review*. Vol. 5, No. 6, November/December 1974.

Parker, Richard A. See Neugebauer.

Petrie, William Flinders. *Amulets*. London: Constable & Company, Ltd., 1914.

*From the Earliest Kings to the XVIth Dynasty*, Vol. 1 of *A History of Egypt*. New York: Charles Scribner's Sons, 1924.

Piankoff, Alexandre. *The Litany of Re*, Vol. 4 of *Egyptian Religious Texts and Representations*. New York: Bollingen Foundation, 1964. Distributed by Pantheon Books.

*The Pyramid of Unas*, Vol. 5 of *Egyptian Religious Texts and Representations*. Princeton: Princeton University Press, 1968.

*The Wandering of the Soul*, Vol. 6 of *Egyptian Religious Texts and Representations.* Princeton: *Princeton University Press, 1974.*

Plato. *Phaedo*. Translated by B. Jowett in *The Works of Plato*. New York: Tudor Publishing Company. No date.

*The Republic*. Translated by Francis Macdonald Cornford. New York and London: Oxford University Press, 1967. First edition published in 1941.

*The Timaeus*. Edited with introduction and notes by R.D. Archer-Hind. New York: Arno Press, 1973. Reprint of 1888 edition.

Plutarch. *'On Exile.'* Translated by Louise Ropes Loomis in *Plutarch — Selected Lives and Essays*. New York: Walter J. Black, Inc., 1971.

*Concerning Such Whom God is Slow to Punish, Of Isis and Osiris*, edited by William W. Goodwin in *Plutarch's Morals*, Vol IV. Boston: Little, Brown, and Company, 1878.

Porphry. *On the Cave of the Nymphs in the Thirteenth Book of the Odyssey*. Translated by Thomas Taylor. London: John M. Watkins, 1917.

Portillo, Jose Lopez. *Quetzalcoatl*. New York: The Seabury Press, 1976.

Pratt, J. Gaither. *Parapsychology: An Insider's View of ESP*. New York: E. P. Dutton & Co., 1966.

Putoff, Harold E. See Targ.

Reisner, George A., and Wm. Stevenson Smith. *History of the Giza Necropolis*. Oxford: Oxford University Press, 1949.

Ritchie, George G. *Return from Tomorrow.* Waco; U.S.A.: Chosen Books, distributed by Word Books, 1978.

Robinson, Herbert Spence, and Knox Wilson. *Myths and Legends of All Nations*. New York: Doubleday & Company, Inc., 1950.

Rogo, D. Scott. *Exploring Psychic Phenomena*. Wheaton, U.S.A.; Madras, India; London: Theosophical Publishing House, 1976.

Ryzl, Milan. *Parapsychology, A Scientific Approach*. New York: Hawthorne Books, Inc., 1970.

Sagan, Carl. *The Dragons of Eden*. New York: Random House, 1977.

Santillana, Giorgio de, and Hertha von Dechend. *Hamlet's Mill*. Boston: Gambit, 1969.

Schafer, Edward H. 'The Sky River.' *Journal of the American Oriental Society*, Vol. 94, No. 4, (1974).

Scholem, Gershom G. *Major Trends in Jewish Mysticism*. New York: Schocken Books, 1946.

Slover, Clark Harris. See Cross.

Smith, Wm. Stevenson. See Reisner.
Spelder, Lynn de. 'Dying in America.' *New Age,* November 1977.
Squire, Charles. *Celtic Myth and Legend*. Van Nuys, U.S.A.: Newcastle Publishing Co. Inc., 1975. Reprint of *The Mythology of the British Islands,* first edition, 1905.
Stahl, William Harris. See Macrobius.
Stecchini, Livio Catullo. 'Notes on the Relation of Ancient Measures to the Great Pyramids', an appendix in *Secrets of the Great Pyramid* by Peter Tompkins. New York: Harper & Row, 1971.
Steiner, Rudolph. *Life Between Death and Rebirth*. New York: Anthroposophic Press, Inc., 1968.
Stevenson, Ian. *Twenty Cases Suggestive of Reincarnation*. New York: American Society for Psychical Research, 1966. (Proceedings of the ASPR, Vol. XXVI, September 1966.)
Swann, Ingo. *To Kiss Earth Good-bye*. New York: Hawthorne Books, Inc., 1975.
Targ, Russell, and Harold E. Putoff. *Mind-Reach: Scientists Look at Psychic Ability*. New York: Delacorte Press/Eleanor Friede, 1977.
Tart, Charles T.'A Psychophysiological Study of Out-of-the-Body Experiences in a Selected Subject.' *Journal of the American Society of Psychical Research,* Vol. 62, No. 1, January 1968.
*PSI: Scientific Studies of the Psychic Realm*. New York: E.P. Dutton, 1977.
Taylor, Peggy and Rick Ingrasci. 'Out of the Body: A New Age Interview with Elizabeth Kübler-Ross.' *New Age,* November 1977.
Thom, Alexander. *Megalithic Sites in Britain*. London: Oxford University Press, 1969.
*Megalithic Lunar Observatories*. London: Oxford University Press, 1971.
Thompson, J. Eric S. *Maya Hieroglyphic Writing,* third edition. Norman: University of Oklahoma Press, 1971.
*Time*. 'Puzzling Out Man's Ascent.' 7 November 1977, Vol. 110, No. 19.
Tompkins, Peter. *Secrets of the Great Pyramid*. New York: Harper & Row, 1971.
*Upanishads*. *The Thirteen Principal Upanishads,* translated by Robert Ernest Hume. London, New York, Toronto: Oxford University Press, 1921.
*The Principal Upanishads,* translated and edited by S. Radhakrishnan. New York: Harper & Brothers, 1953.
Vassar, Carol. See Palmer.
Virgil. *The Georgics*. Translated by J. W. Mackail in *Virgil's Works*. New York: Random House, 1950.
*The Aeneid*. Translated by W.F. Jackson Knight. Harmondsworth, England; Baltimore, U.S.A.; Ringwood, Victoria, Australia: Penguin Books, 1965.
Wainwright, G.A. 'The Origin of Storm-Gods in Egypt.' *The Journal of Egyptian Archaeology*. Vol. 49 (1963).
*The Sky-Religion In Egypt*. Westport, U.S.A.: Greenwood Press, 1971. Reprint of 1938 edition.
Waters, Frank. *The Book of the Hopi*. New York: Viking, 1963.
Weiss, Jess E. *The Vestibule*. Port Washington, U.S.A.: Ashley Books, Inc., 1972.
White, John, ed. *Frontiers of Consciousness*. New York: The Julian Press, Inc., 1974.
Wiedemann, Alfred. *The Ancient Egyptian Doctrine of the Immortality of the Soul*. New York: G. P. Putnam's Sons; London: H. Grevel & Co., 1895.
Willoughby, Harold R. *Pagan Regeneration*. Chicago: The University of Chicago Press, 1960. Also University of Toronto Press, 1960. Reprint of 1929 edition.

Wilson, Knox. See Robinson.
Woodward, Kenneth L., and Rachel Mark. 'Life After Life?' *Newsweek*, 1 May 1978, Vol. XCI, No. 18.
*The Zohar*. Vol. 1. Translated by Harry Sperling and Maurice Simon. London, Jerusalem, New York: The Soncino Press, 1934, first published. Reprinted in 1970.
Zubeck, John P., ed. *Sensory Deprivation: Fifteen Years of Research*. New York: Appleton-Century Crofts (Educational Division), Meredith Corporation, 1969.

# PICTURE CREDITS

Akademische Druck und Verlagsanstalt, Graz *25*. G. Bell & Sons, London *45, 52 above, 60*. British Crown Copyright, reproduced with permission of the Controller of Her Britannic Majesty's Stationery Office *11 above*. British Museum, reproduced by courtesy of the Trustees *189*. Brown University Press, Providence, Rhode Island: *Egyptian Astronomical Texts*, O. Neugebauer and R. Parker *80, 82, 86 above*. *Canadian Journal of Psychology*, 1966, *20*, pp. 316-336, Zubek and MacNeil *140*. Cassell & Co., London and New York *48 above, 51, 53, 56, 62, 64, 168*. Constable & Co., London *38 above*. Thos. Y. Crowell Inc., New York *57*. Dover Publications Inc., New York *30*. Wm. R. Fix *11 below, 14, 15, 16, 42, 43, 47, 48 below, 49, 58 above, 73, 78-79* (after Epigraphic Survey, Oriental Institute plates), *85 below, 86 below* (after Epigraphic Survey, Oriental Institute plates), *90, 94, 97* (after R. H. Coleman), *101 above* (after Selim Hassan), *102 above, 103, 129* (after R. Cook), *139, 144, 145, 146, 147, 148, 149, 150, 181*. Freer Art Gallery, Washington *115*. Eleanor Friede, New York *112*. Gambit, Ipswich, Mass. *20, 21, 27 below*. Gambit/Stefan Fuchs, University of Frankfurt *23*. Government Press, Cairo *101 below*. Hale Observatories, Pasadena, Cal. *156, 204, 206, 217*. Hamlyn Group, Feltham, Middlesex *24, 136*. Charles Hapgood, Winchester, N.H. *13 below, 22 above*. Harvard College Observatory, Cambridge Mass. *154*. Hutchinson Publishing Group, London *208*. Institut d'Ethnologie, Paris *27 above*. Jet Propulsion Laboratory, Pasadena, Cal. *109*. Diannellen Knox *34 above, 34 below* (after E. A. W. Budge), *67 below* (after Selim Hassan), *165 above* (after E. A. W. Budge), *226* (after W. K. C. Guthrie), *230*. Kunsthistorisches Museum, Vienna *96*. Ernst and Johanna Lehner, *Lore and Lure of Outer Space*, Tudor, New York *63, 158, 184, 196, 202, 232, 242*. *Folklore and Symbolism of Flowers, Plants and Trees 118*. Richard Lepsius *84 above*. Library of Congress, Washington *13 above* (reproduced from an engraved map in the collections of the Geography and Map Division), *152*. Lick Observatory, San Jose, Cal. *22 below, 179, 228*. Little, Brown & Co., Boston *155*. J. Norman Lockyer *59 above*. R. N. Lockyer *81*. Macmillan Co., New York *75*. Medford, Oregon, Schools Planetarium *214*. Merseyside County Museums, Liverpool *223 above*. Methuen, London *186*. The Metropolitan Museum of Art, New York *84 below, 117, 135* (Museum excavations 1923-24; Rogers Fund 1925). Middendorf, Peru III *54*. M.I.T. Press, Boston *58 below*. Mt Wilson and Palomar Observatories, Pasadena, Cal. *199*. Museum of Man, Paris *198*. National Library, Paris *133*. National Museum of Denmark, Department of Ethnology, Copenhagen *126*. Open Court/Methuen, La Salle, Ill.: *The Gods of the Egyptians*, E. A. W. Budge *67 above, 68, 69, 71 above, 95*. Oriental Institute, Chicago *81 above, 85 above, 100*. Philbrook Art Center, Tulsa, Okla. *137, 142*. Princeton University Press, Princeton, N.J. *82, 92, 125, 143, 155, 160*. G. P. Putman's Sons, New York *170*. Robinson & Watkins, London: *Qabbalah*, Isaac Myer *192*. Routledge & Kegan Paul Ltd., London *33, 35, 37, 38 below, 39, 70, 71 below, 76, 77, 98, 102 below, 162, 172, 174*. Charles Scribner's Sons, New York *165 below, 166*. Smithsonian Institution, Washington *26*. Rudolf Steiner Press, London *151*. Thames & Hudson, London: *The Tree of Life*, Roger Cook *119, 128, 190*. Times Books, New York: *America B.C.*, Barry Fell, © 1976 by Barry Fell *31*. Gibson L. Towns *61*. West Baffin Eskimo Co-operative Limited, Cape Dorset, N.W.T. *130*. D. D. Zink *55*.

Although every effort has been made to ensure that permission for all material was obtained, those sources not formally acknowledged will be included in all future editions of this book.

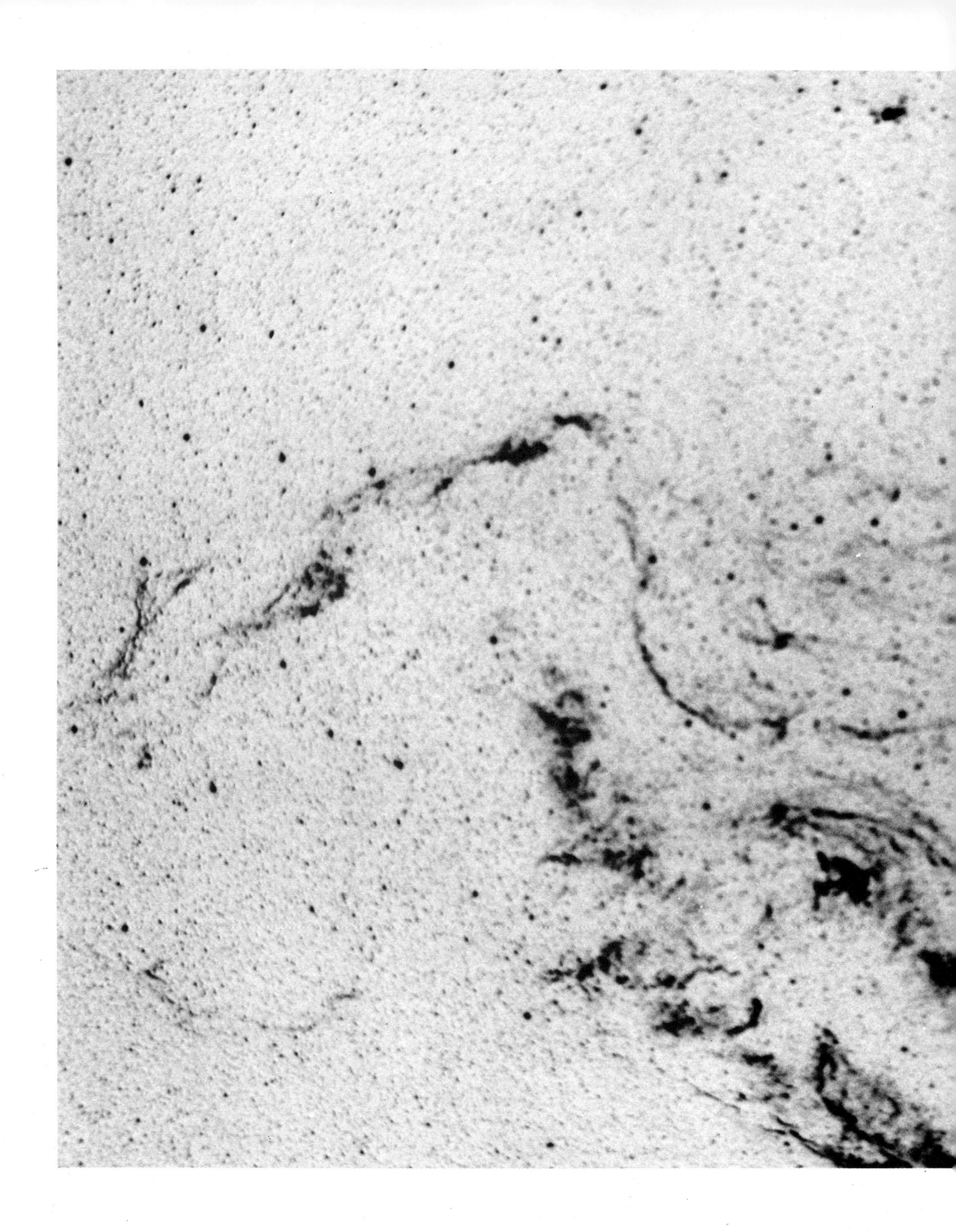

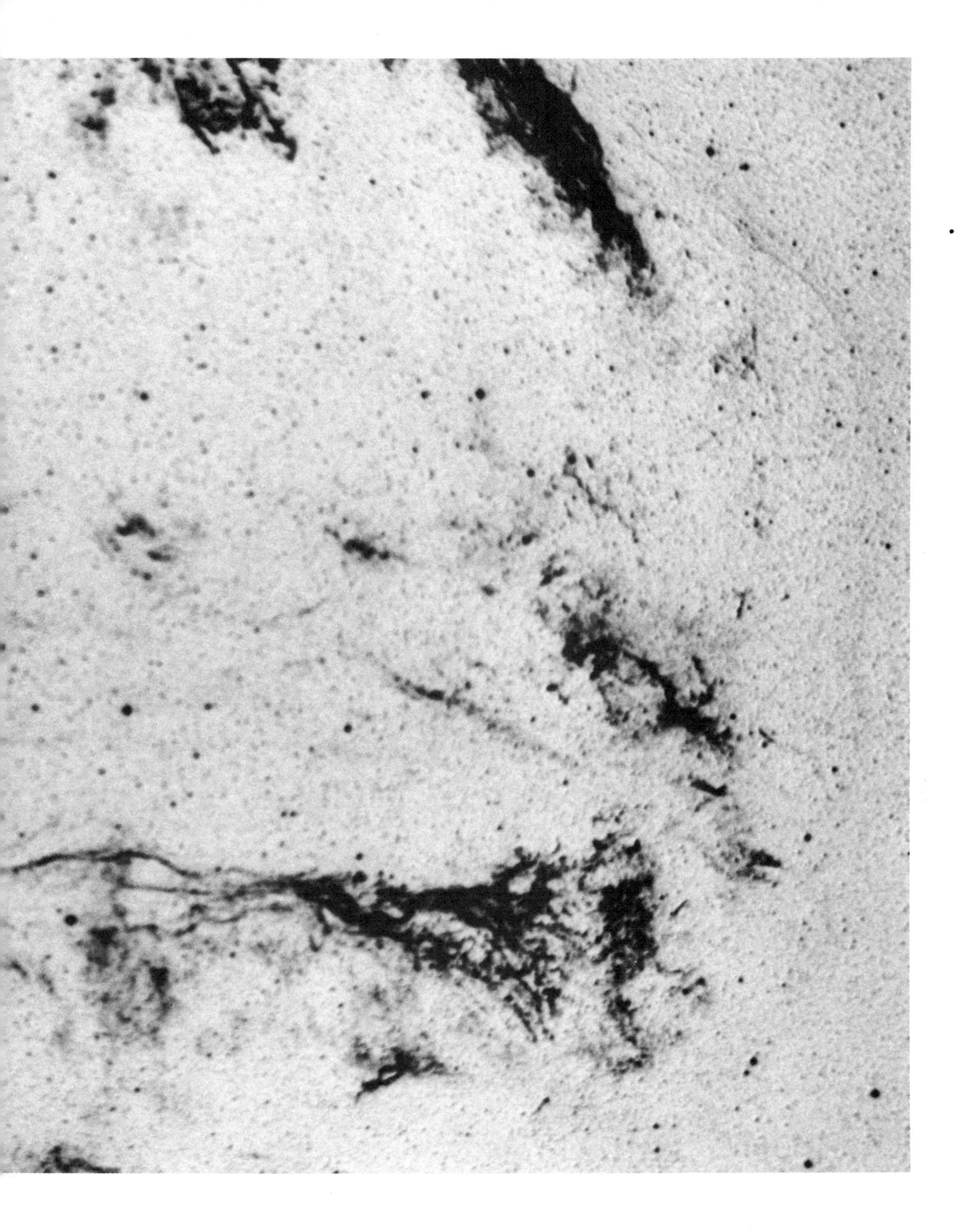